ADVANCES IN EXPERIMENTAL MEDICINE AND BIOLOGY

Volume 782

AF180803

For further volumes:
http://www.springer.com/series/5584

Michael J. Richardson • Michael A. Riley
Kevin Shockley
Editors

Progress in Motor Control

Neural, Computational and Dynamic Approaches

Springer

Editors
Michael J. Richardson
University of Cincinnati
Cincinnati, Ohio
USA

Michael A. Riley
University of Cincinnati
Cincinnati, Ohio
USA

Kevin Shockley
University of Cincinnati
Cincinnati, Ohio
USA

ISSN 0065-2598
ISBN 978-1-4899-9644-2 ISBN 978-1-4614-5465-6 (eBook)
DOI 10.1007/978-1-4614-5465-6
Springer Dordrecht Heidelberg London New York

Printed on acid-free paper

Springer is part of Springer Science+Business Media (www.springer.com)

Preface

This volume is the most recent installment of the *Progress in Motor Control* series. It contains contributions based on presentations by invited speakers at the Progress in Motor Control VIII meeting held in Cincinnati, OH, USA in July, 2011. Progress in Motor Control is the official scientific meeting of the International Society of Motor Control (ISMC). There were 23 invited presentations at the meeting, which was organized into eight themed symposia and included a special ISMC Past President's Address by Michael Turvey, along with 137 poster presentations.

The Progress in Motor Control VIII meeting, and consequently this volume, were meant to provide a broad perspective on the latest research on motor control in humans and other species. The invited talks at the meeting addressed topics such as neural regeneration, the mirror neuron system, movement disorders, dynamical systems models and analyses, cortical representation and control of movement, spinal circuitry for movement control, neuromechanics, motor learning, computational modeling, and interactions between cognitive and motor processes. Neuroscience, psychology, physiology, kinesiology, biomechanics, engineering, neurology, physics and applied mathematics are among the disciplines represented by the chapters and their authors. The chapters also reflect a broad range of approaches and theoretical points of view, including neural, computational, and dynamical systems perspectives.

This diversity of perspectives and approaches, while certainly not exhaustive or even fully representative, provides a flavor of the complex and multi-faceted nature of motor coordination and control. While it is clear that much progress has been made—fueled in part, hopefully, by the eight Progress in Motor Control meetings to date and the publications associated with them—it is nonetheless apparent that a thorough and complete understanding of motor control is not yet within our grasp. It will require a sustained effort to achieve this understanding, and continued efforts to synthesize the results of studies that are accruing at what seems to be an exponentially increasing rate. We hope that this volume contributes to these important goals in at least some small way.

We would like to acknowledge the extremely valuable help of Jamie Miller and the University Conferencing staff who helped us plan and execute Progress in Motor Control VIII. Thanks are also due to the graduate students from the Perceptual-Motor Dynamics Laboratory at the Center for Cognition, Action, and Perception in the University of Cincinnati Psychology Department who helped make the meeting run smoothly—Dilip Athreya, Laura Bachus, Scott Bonnette, Tehran Davis, Nikita Kuznetsov, MaryLauren Malone, Michael Tolston, Julie Weast-Knapp, and Eli White. The meeting was supported financially by the National Institute of Neurological Disorders and Stroke (grant number 1R13NS073205-1). We also appreciate the input from past and present ISMC officers and previous Progress in Motor Control organizers. Finally, we would like to thank Arthur Smilios at Springer for his encouragement and assistance with putting together this volume.

Cincinnati, Ohio, USA
April 2012

Michael J. Richardson
Michael A. Riley
Kevin Shockley

Contents

Contributors

Ramesh Balasubramaniam Sensorimotor Neuroscience Laboratory, McMaster University, Hamilton, ON L8S 4K1, Canada
e-mail: ramesh@mcmaster.ca

Jessica A. Bernard Department of Psychology, University of Michigan, 401 Washtenaw Avenue, Ann Arbor, MI 48109-2214, USA
e-mail: jessbern@umich.edu

Numa Dancause Groupe de Recherche sur le Système Nerveux Central (GRSNC), Département de Physiologie, Pavillon Paul-G-Desmarais, Université de Montréal, 2960, Chemin de la Tour, bureau 4138, Montréal, Québec H3T 1J4, Canada
e-mail: numa.dancause@umontreal.ca

Didier Delignières EA 2991 Movement to Health - Euromov, UFR STAPS, University Montpellier 1, 700 avenue du Pic Saint Loup, 34090 Montpellier, France
e-mail: didier.delignieres@univ-montp1.fr

Brett W. Fling School of Kinesiology, University of Michigan, 401 Washtenaw Avenue, Ann Arbor, MI 48109-2214, USA
e-mail: bfling@hs.uci.edu

Leonardo Fogassi Dipartimento di Neuroscienze e Istituto Italiano di Tecnologia (RTM), Università di Parma, Parma, Italy

Dipartimento di Psicologia, Università di Parma, Parma, Italy
e-mail: leonardo.fogassi@unipr.it

Leslie A. Gilmore Department of Biological Sciences, Northern Arizona University, Flagstaff, AZ 86011-5640, USA

Adrian M. Haith Department of Neurology, Johns Hopkins University, Baltimore, MD 21205, USA
e-mail: adrian.haith@jhu.edu

John W. Krakauer Departments of Neurology and Neuroscience, Johns Hopkins University, Baltimore, MD 21205, USA
e-mail: jkrakau1@jhmi.edu

Robert G. Kalb Department of Neurology, Children's Hospital of Philadelphia, Perelman School of Medicine, University of Pennsylvania, 3615 Civic Center Boulevard, Philadelphia, PA 19104, USA

Research Institute and Division of Neurology, Department of Pediatrics, Children's Hospital of Philadelphia, Perelman School of Medicine, University of Pennsylvania, 3615 Civic Center Boulevard, Philadelphia, PA 19104, USA
e-mail: kalb@email.chop.edu

Youngbin Kwak Neuroscience Program, University of Michigan, 401 Washtenaw Avenue, Ann Arbor, MI 48109-2214, USA

Center for Cognitive Neuroscience, Duke University, Durham, NC 27708, USA
e-mail: youngbin.kwak@duke.edu

Stan L. Lindstedt Department of Biological Sciences, Northern Arizona University, Flagstaff, AZ 86011-5640, USA

Vivien Marmelat EA 2991 Movement to Health - Euromov, UFR STAPS, University Montpellier 1, 700 avenue du Pic Saint Loup, 34090 Montpellier, France

John G. Milton W. M. Keck Science Department, The Claremont Colleges, Claremont, CA 91711, USA
e-mail: jmilton@jsd.claremont.edu

Jenna A. Monroy Department of Biological Sciences, Northern Arizona University, Flagstaff, AZ 86011-5640, USA

Kiisa C. Nishikawa Department of Biological Sciences, Northern Arizona University, Flagstaff, AZ 86011-5640, USA
e-mail: Kiisa.Nishikawa@nau.edu

Krysta L. Powers Department of Biological Sciences, Northern Arizona University, Flagstaff, AZ 86011-5640, USA

Rachael D. Seidler Department of Psychology and School of Kinesiology, University of Michigan, 401 Washtenaw Avenue, Ann Arbor, MI 48109-2214, USA
e-mail: rseidler@umich.edu

Luciano Simone Dipartimento di Psicologia, Università di Parma, Parma, Italy

Theodore A. Uyeno Department of Biology, Valdosta State University, Valdosta, GA 31698-0015, USA

Lei Zhang Research Institute and Division of Neurology, Department of Pediatrics, Children's Hospital of Philadelphia, Perelman School of Medicine, University of Pennsylvania, 3615 Civic Center Boulevard, Philadelphia, PA 19104, USA
e-mail: ZhangL@email.chop.edu

Weiguo Zhou Research Institute and Division of Neurology, Department of Pediatrics, Children's Hospital of Philadelphia, Perelman School of Medicine, University of Pennsylvania, 3615 Civic Center Boulevard, Philadelphia, PA 19104, USA
e-mail: weiguozhou@gmail.com

Chapter 1
Model-Based and Model-Free Mechanisms of Human Motor Learning

Adrian M. Haith and John W. Krakauer

Introduction

In laboratory settings, motor learning has typically been studied in the context of adaptation paradigms in which subjects must learn to compensate for a systematic perturbation—either some manipulation of visual feedback (Krakauer et al. 2000) or a change in the dynamics of the motor apparatus, e.g., a force applied to the hand (Shadmehr and Mussa-Ivaldi 1994), Coriolis forces induced by rotation of the body (Lackner and Dizio 1994), or an inertial load attached to the arm (Krakauer et al. 1999). What is typically observed in these tasks is a monotonic improvement in performance that is initially rapid, and then slows to an asymptote close to initial baseline levels of performance. The progress of learning is well described by exponential fits, implying that the amount of improvement on each trial is proportional to the error (Thoroughman and Shadmehr 2000; Donchin et al. 2003). This kind of fast, trial-by-trial reduction in systematic errors is typically referred to as *adaptation*. The term adaptation has been used in some cases to imply a particular mechanism of learning; however, we will adhere to a behavioral definition (as a gradual reduction in error following an abrupt change in conditions) and describe potential underlying learning mechanisms in more computational terms. As we will argue, learning in adaptation paradigms is likely predominantly mediated by a specific learning mechanism that is based on changing an internal forward model.

Not all motor learning falls under our behavioral definition of adaptation. Often one learns to synthesize entirely novel movements even when there is no perturbation, e.g., learning to swing a golf club, hit a tennis serve, balance a pole, or drive a car. Although this kind of learning corresponds more closely to everyday usage of

A. M. Haith (✉)
Department of Neurology, Johns Hopkins University, Baltimore, MD 21205, USA
e-mail: adrian.haith@jhu.edu

J. W. Krakauer
Departments of Neurology and Neuroscience, Johns Hopkins University,
Baltimore, MD 21205, USA
e-mail: jkrakau1@jhmi.edu

M. J. Richardson et al. (eds.), *Progress in Motor Control*,
Advances in Experimental Medicine and Biology 782,
DOI 10.1007/978-1-4614-5465-6_1, © Springer Science+Business Media New York 2013

the term "motor learning," it has hardly been studied in laboratory settings. The few exceptions typically involve learning to manipulate an unfamiliar, possibly complex virtual object (Carmena et al. 2003; Mosier et al. 2005; Nagengast et al. 2009; Sternad et al. 2011). In these kinds of tasks, subjects progress from initial incompetence to a high degree of proficiency, even approaching theoretically optimal behavior. However, performance improvements are far slower than in adaptation paradigms: while tens of trials are usually enough to reach asymptote after a systematic perturbation is introduced, performance in these more complex tasks continues to improve over hundreds of trials and even across days. This slow improvement is not entirely due to the unfamiliarity of the task. Even in much more simple tasks that involve maneuvering a cursor along a constrained path (Shmuelof et al. 2012) or through a series of via points (Reis et al. 2009), overall variability in task performance reduces substantially over days of practice, even though subjects immediately exhibit near-perfect performance at slow speeds. It appears that a qualitatively different kind of learning may be occurring in these tasks—one that is not reliant on compensating for the highly salient errors that are present in adaptation settings, but instead is associated with incrementally improving the quality of one's movements with practice. We define this long-term reduction in movement variability as *skill learning*. It is not currently clear whether adaptation, skill, and learning to control external objects draw upon identical, overlapping, or entirely different neural mechanisms.

In this review, we argue for the existence of two distinct mechanisms underlying motor learning: (1) a model-based system in which improvements in motor performance occur indirectly, guided by an internal forward model of the environment which is updated based on prediction errors, and (2) a direct, model-free system in which learning occurs directly at the level of the controller and is driven by reinforcement of successful actions. These distinct learning systems are each suited to different tasks and as such are complementary to one another. Model-based processes are likely to predominate in adaptation paradigms, and model-free processes predominate in skill tasks. However, we argue that both can contribute to learning in any given task.

Theory: Model-Based and Model-Free Approaches to Learning Control Policies

We adopt a general definition of motor learning as the process of improvement in execution of a task according to some chosen measure of performance such as increased chance of success or decreased effort (or potentially a combination of the two). Formally, we describe the state of learning in terms of a control policy π mapping current states, stimuli and time to motor commands u_t,

$$u_t = \pi(x_t, s_t, t).$$

This general framework can encompass multiple levels of description. A control policy could describe selection of a single action per trial or describe an ongoing stream of motor commands in continuous time according to the instantaneous state. The motor commands u_t could model a high-level decision such as which direction to move the hand or a low-level decision such as which muscles to activate and when. The stimulus s_t would typically correspond to an observed target location and the state x_t would reflect the state of the motor plant. Any systematic, experience-driven change in this control policy can be described as motor learning. The quality of each potential control policy can be quantified in terms of the expected outcome value, i.e., the average performance that would be expected to be obtained when following that control policy for a given task. In studying motor learning, we study the process whereby individuals use experience to improve their control policy.

The optimal policy will depend on two specific things: (1) the structure of the task, i.e., which states are associated with valuable or successful outcomes and what costs may be associated with different states or actions and (2) the dynamics of the motor apparatus and environment, i.e., how do motor commands affect the state. In most motor control paradigms we would generally expect that the structure of the task is unambiguous; however, in general it may be that neither the task structure nor the dynamics dynamics is known precisely.

This general framework and the problem of determining suitable actions in an uncertain environment based on ongoing observations is precisely the theoretical problem studied, at a more abstract level, in the field of *reinforcement learning* (Sutton and Barto 1998). At the heart of reinforcement learning theory is the notion of the *value function* $V(x_t, t)$ which reflects, for a given control policy, the total future reward that can be expected to be gained given the current state and time. The goal of reinforcement learning is to determine the optimal value function—from which the optimal policy follows straightforwardly.

Different variants of reinforcement learning differ in exactly what kind of value function is represented and how this value function is updated based on experience. In particular, two distinct computational strategies have emerged for using experience to update estimated values and thereby determine optimal control policies. The first approach is to use experience to build models of the dynamics of the motor apparatus and environment and the structure of the task, and compute the value function based on these models (Fig. 1.1a). This approach is termed *model-based learning*. Note that model-based learning of this kind is very different from what most people understand intuitively by the term 'reinforcement learning.'

A second approach, which accords with most people's informal or colloquial use of the term 'reinforcement learning,' is to learn the value function directly through a process of trial and error—explore the space of potential actions in each state and keep track of which states and actions lead, either directly or indirectly, to successful outcomes (Fig. 1.1b). This approach is often termed *model-free*, in contrast to model-based approaches. Other learning strategies are clearly possible besides the model-free and model-based approaches described here. However, these represent the most common approaches.

While model-free strategies clearly work and can in certain cases be shown theoretically to be guaranteed to converge upon optimal behavior (Sutton and Barto 1998), learning by trial and error is typically very slow in terms of the number of attempts necessary before a good policy can be acquired, even in relatively simple

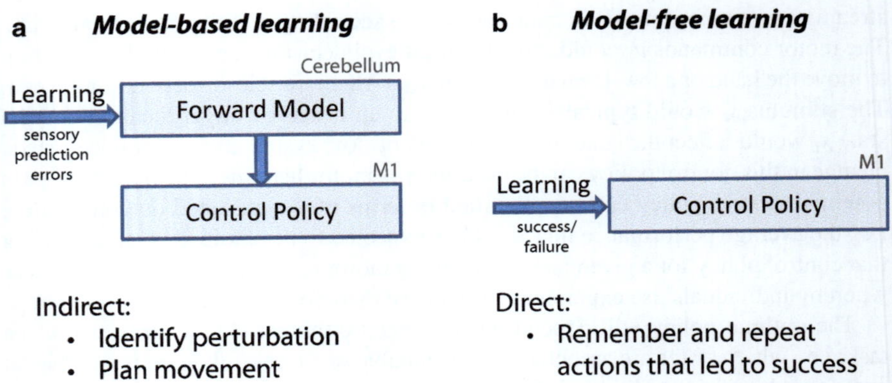

Fig. 1.1 Comparison of model-based and model-free strategies for updating a control policy based on experience. **a** Model-based learning schematic. Changes to the control policy are brought about indirectly through first updating a forward model of the motor apparatus and environment based on sensory prediction errors, then using this knowledge to calculate an appropriate controller for the current task. **b** Model-free learning schematic. The control policy is updated directly based on reward prediction errors

environments. Model-based learning, by contrast, makes the best possible use of all observations. Any information acquired about the outcome of a particular action is retained and can influence planning of future movements, regardless of whether that action led to success or not. Model-based methods also allow more principled generalization. If the reward structure of the task changes in a known way (e.g., the target moves to a new location), an appropriate new control policy can be computed based on the model of the dynamics that was built in the context of the previous task.

The major disadvantage with a model-based approach is that although the value of any state/action pair can in principle be computed exactly, it can be prohibitively computationally intensive to do so. Existing methods for computing the optimal policy typically involve either dynamic programming—a backward iteration through time to exhaustively compare all possible paths to the target and identify the best ones, or some iterative sequence of approximations to the value function or policy that converge upon a local optimum. Details of these methods are beyond the scope of this chapter (but for excellent introductions see Bertsekas 1996; Sutton and Barto 1998; Todorov 2007).

The complexity associated with computing optimal value functions and policies need not preclude biological systems from utilizing some form of model-based control. In certain very simple scenarios, it can be trivial to compute the optimal policy given a specification of the task and plant e.g., if the action on a given trial is simply the aiming direction for a particular movement, then a model-based solution to a rotational perturbation simply amounts to subtracting an estimate of the rotation angle from an observed target angle. The feasibility of the model-based approach therefore largely depends on the nature of the task. Even if the computations are simple, however, errors may still arise from accumulation of noise that inevitably accompanies computations in biological systems (McGuire and Sabes 2009).

Model-free approaches, by contrast, require only relatively trivial computations because experiences lead directly to changes in the controller. Unlike a model-based approach, there is no intermediate forward model representation and no calculation required to transform a forward model into a control policy. In the long-run, model-free approaches tend to deliver superior performance on a particular task because they do not rely so heavily on noisy computations each and every time a movement must be made. The disadvantage is that the scope of the learned control policy is restricted to the task performed during learning. Even if the reward structure of the task changes in a known way, one must start from scratch (or at least from some previous but incorrect control policy). This is in sharp contrast to the flexibility offered by model-based learning.

In summary, if one wants to learn a good control policy in an uncertain environment, a model-based learning strategy is, in principle, the most powerful and flexible approach but requires unwieldy computations. Direct, model-free approaches rely only on simple computations but can require far more training (exploration) before they lead to a competitive policy. What learning strategy do animals use when placed in a situation where they must learn what to do? The contrasting ways in which model-based and model-free learning mechanisms should be expected to generalize to novel scenarios can act as hallmarks that potentially allow us dissociate an animal's learning strategy based on observing its behavior.

Model-Based and Model-Free Learning in Operant Conditioning Paradigms

In a situation where an animal must learn what actions will lead to reward, such as a rat navigating through a maze to find food, it seems that animals adopt both model-based and model-free learning mechanisms in parallel (Daw et al. 2005). Although any given control policy could be arrived at by either model-based or model-free strategies, these two modes of control can be dissociated by changing the reward structure of the task. In rodents this is typically achieved by stimulus revaluation. For instance, imagine examining the behavior of thirsty rats in a maze that they had learned while they were hungry and seeking food. Under a model-free approach, the thirsty rat will have no way of knowing how to obtain water and will likely either behave like a naïve rat, or rely on the same policy that led to reward while hungry. A model-based approach, by contrast, will enable the rat to flexibly change its behavior immediately in line with its new objective of finding water instead of food (provided, of course that it had previously explored the maze sufficiently to have found the location of the water). In practice, rewards are typically revalued either by sating the animal prior to the task or, more drastically, pairing a familiar food with a strongly aversive stimulus (e.g., poison).

Behavior in such devaluation paradigms has been studied extensively, leading to a classical division between *goal-directed* behavior, in which animals are sensitive to reward devaluation, and *habitual* behavior in which they are not (Killcross and

Coutureau 2003; Balleine and O'Doherty 2010; van der Meer and Redish 2011). Behavior tends to be goal-directed early in learning but becomes more habitual later on (Balleine and O'Doherty 2010). These differences in behavior can be interpreted in terms of reinforcement learning: goal-directed behavior can be understood as model-based, while habitual behavior is model-free (Daw et al. 2005; Dayan 2009). The transition from goal-directed to habitual with experience can even be explained as an evolving, intelligent trade-off between the advantages of each strategy.

Remarkably, these alternative model-based and model-free strategies are neurally dissociable. Lesions to distinct regions of prefrontal cortex can isolate one pattern of behavior or another in hungry rats (Balleine and Dickinson 1998; Killcross and Coutureau 2003). Sequential decision-making tasks in humans have revealed that their behavior can similarly be decomposed into model-based and model-free components (Fermin et al. 2010; Gläscher et al. 2010), while fMRI reveals that these components have distinct underlying neural substrates (Gläscher et al. 2010).

The kinds of control tasks that we are primarily concerned with in this review are quite different from the problem that a rat faces in a maze. In decision-making tasks it is the high-level choice of which path to follow at a junction that is of interest. The low-level movements that register this decision are considered incidental. In motor control, however, it is precisely these low-level movements that are of interest. Critically, control of movements can be cast within the same broad theoretical framework used to describe decision-making. The only differences are that movements of the eyes and limbs occupy a space of potential states and actions that is continuous and potentially high-dimensional, and decisions must be made in continuous time. Nevertheless, the same considerations for solving the general problem apply as in more discrete domains. In particular, both model-based and model-free learning strategies are possible and have similar advantages and disadvantages as in discrete domains. We will argue that, as in the case of rodent decision-making, both strategies are employed by the motor system for continuous control of movement. The underlying neurophysiology may, however, be quite different for the motor system as compared with the discrete action selection paradigms studied in rodents.

Model-Based Motor Learning

Forward models—neural networks which generate predictions about future states of the motor system given a current state and an outgoing motor command—have long been posited to be utilized by the motor system (Wolpert and Miall 1996). Model-based learning has become a dominant framework for understanding human motor learning, with arguably the majority of theories of motor learning assuming a model-based perspective (Shadmehr and Krakauer 2008; Shadmehr et al. 2010). The proposed advantages of maintaining a forward model are twofold: (1) A forward model allows for faster and more precise estimation of the state of the body and/or environment, and (2) Forward models may aid in planning future movements by directing changes in the controller itself, i.e. they may participate in model-based control. While (1) has by now become a relatively uncontroversial claim, (2) is much more difficult to establish.

Before assessing the case for model-based learning in the motor system, we briefly disambiguate model-based learning from learning involving *inverse models*. The simplest kind of controller considered in motor learning theories is a static mapping from a desired outcome to a single action. Such controllers have been referred to as *inverse models* since they are the direct inverse of the forward model. However, inverse model are not really "models" in the true sense of the word—they do not provide an internal representation of any process occurring in the outside world. It is more accurate to think of inverse models as simple control policies. An inverse model control policy can be arrived at in a model-based manner by first learning a forward model and then inverting it (Jordan 1992). Alternatively, changes to an inverse model could be driven directly by task errors (Thoroughman and Shadmehr 2000). We would not describe such learning as model-based, however, since the learning occurs directly at the level of the controller rather than via a forward model representation of the task or plant. Learning of this kind is only really feasible in simple, single-time step scenarios.

Theories based on the notion of inverse models are fairly limited in scope. More generally, motor control is described in terms of time-dependent feedback control policies (Todorov and Jordan 2002). In this context, there is no way to directly update the control policy based on performance errors. By contrast, model-based learning is a very general approach to obtaining a good control policy that is applicable to any problem that can be framed as a Markov decision process. The only limitation to model-based learning is being able to gather enough information to build the model.

Nothing is presently known about the neural computations that underlie the translation of knowledge about the environment in the form of a forward model into a control policy. However, even though the potential mechanisms underlying model-based control processes are poorly understood, this understanding is not necessary to establish whether or not it occurs. Here, we focus on reviewing the evidence at the behavioral level for the existence of forward models and their involvement in motor learning.

The Cerebellum and Forward Models

The cerebellum has long been implicated in motor control and coordination and has emerged as the most likely neural substrate of putative internal models (Bastian 2006; Shadmehr and Krakauer 2008; Wolpert et al. 1998). Patients with hereditary cerebellar ataxia or lesions to the cerebellum have general difficulties in coordinating movement and are grossly impaired in adaptation tasks (Martin et al. 1996; Maschke et al. 2004; Smith and Shadmehr 2005; Tseng et al. 2007; Synofzik et al. 2008; Rabe et al. 2009; Criscimagna-Hemminger et al. 2010; Donchin et al. 2011). There are many potential roles for the cerebellum in learning that might give rise to such an adaptation deficit in cerebellar ataxia. The cerebellum may, for instance, compute an inverse model that directly maps desired outcomes to actions (Medina 2011). We argue here, however, that the adaptation deficit following cerebellar damage stems from an inability to learn forward models.

Neurophysiological recordings from the cerebellum show that Purkinje cell simple spike activity reflects the kinematics of movement, and not the motor commands required to achieve the kinematics (Pasalar et al. 2006). This finding clearly demonstrates that the output of the cerebellum is not directly related to motor output, as would be predicted if the cerebellum were computing an inverse model or otherwise contributing directly to control. Furthermore, Purkinje cell activity during movement precedes the actual kinematic state of the limb (Roitman et al. 2005). So this activity in the cerebellum does not simply reflect a reporting of sensory feedback—instead it appears that the cerebellum implements an internal forward model that predicts the kinematic or sensory consequences of motor commands before that information actually becomes available from the periphery.

Numerous studies have argued from a behavioral standpoint that an estimate of state from a forward model underlies state-specific feedback corrections during movement (Ariff et al. 2002; Chen-Harris et al. 2008; Wagner and Smith 2008; Munuera et al. 2009). This process appears to be cerebellar-dependent (Miall et al. 2007; Xu-Wilson et al. 2009). Together with the above-mentioned neurophysiological findings, these studies make a strong case that the cerebellum generates predictions about future motor states on the basis of outgoing motor commands, and that these predicted states are made available to an already-learned feedback controller that guides ongoing execution of a movement. While this constitutes model-based control of sorts (Mehta and Schaal 2002), in this article we are more interested in the question of whether a forward model brings about changes in the controller, rather than influencing control only through estimates of state. Nevertheless, if forward models exist and can be used to guide online feedback control, it perhaps makes it more likely that the same forward models might participate in planning feedforward control.

Evidence for Forward Model Involvement in Feedforward Control

An often-cited instance where predictions of a forward model are claimed to influence feedforward control (as opposed to only feedback control) is in compensating for the consequences of one effector's actions on another—for instance stabilizing one hand holding a load while removing that load with the other hand or increasing grip force on an object to prevent slippage when accelerating it upward. It has often been argued that such *anticipatory control* is possible because of a forward model that predicts adverse consequences of an upcoming action before it has happened, enabling an appropriate compensation to be planned and executed concomitantly (Wolpert and Miall 1996; Flanagan and Wing 1997; Wolpert et al. 2011). Although the use of a forward model could, in principle, enable this kind of anticipatory control, coordination *per se* is no proof of the existence of forward models. Anticipatory control is simply a feature of a good control policy and there is no way of knowing how this controller may have been arrived at simply by observing it in action. Good coordination could have been learned via model-free mechanisms through trial and error.

Studies of anticipatory control in cerebellar ataxia patients offer some clues as to the nature of anticipatory control. Interestingly, cerebellar ataxic patients demonstrate intact coordination in manual unloading tasks (Diedrichsen et al. 2005) and exhibit intact modulation of grip force with varying load forces (although baseline grip forces are abnormal) (Rost et al. 2005), suggesting that forward models are not at all a prerequisite for performing coordinated movement. Cerebellar patients do, however, show impairment in learning novel anticipatory adjustments (Nowak et al. 2004; Diedrichsen et al. 2005). This suggests that initial acquisition of anticipatory control is facilitated by a forward model that can predict the consequences of the actions of one effector on the goals of another but, with prolonged practice, coordinated control eventually becomes independent of the forward model. To put it another way, there may be a transition from model-based to model-free mechanisms.

The notion of model-based learning implies that improvements in performance are driven by errors in the prediction of a forward model. One plausible alternative to this idea is that adaptation is driven by the feedback corrections one makes to correct errors, rather than by the errors themselves (Kawato and Gomi 1992). This does not appear to be the case for reaching movements, however: learning rates in adaptation tasks are identical whether or not feedback corrections are allowed during movement (Tseng et al. 2007). Similarly, corrective saccades do not appear to be necessary to adapt saccade amplitude (Wallman and Fuchs 1998).

Although not driven by corrective movements, adaptation may not necessarily be driven by prediction errors of a forward model. If control is mediated by an inverse model, changes to a control policy could be driven directly by task errors, without any need for a forward model. In most cases, task errors and prediction errors are closely aligned. In certain cases, however, performance errors and prediction errors can be dissociated. For instance, saccades to visual targets usually tend to fall slightly short of the target, but this shortfall does not lead to an increase in saccadic gain as one would expect if it were induced through a target jump. In fact, if the target is surreptitiously jumped mid-saccade such that the eye lands perfectly on the target every time, then saccadic gain actually begins to decrease despite the absence of performance errors (Wong and Shelhamer 2011). Indeed it is even possible for adaptation to occur in the opposite direction from a task error. This provides compelling evidence that prediction errors and not task errors are what drive motor adaptation.

A similar, even more striking result can be found for reaching movements. In a study by Mazzoni and Krakauer (2006) (Fig. 1.2a), subjects were exposed to a 45° rotation of visual feedback but were also provided an explicit strategy to counter the rotation: simply aim to an adjacent target deliberately spaced at a 45° separation from the true target. Initially, subjects were able to flawlessly implement the strategy and hit the target. However, performance rapidly began to drift away from the target in the direction of the perturbation despite the fact that the task was being performed without errors. It therefore does not seem to be task error *per se* that drives adaptation, but discrepancies between predicted and observed behavior. Interestingly, this drift effect does not persist indefinitely—after prolonged exposure, subjects begin to reduce their errors again, suggesting that there is some component of learning that

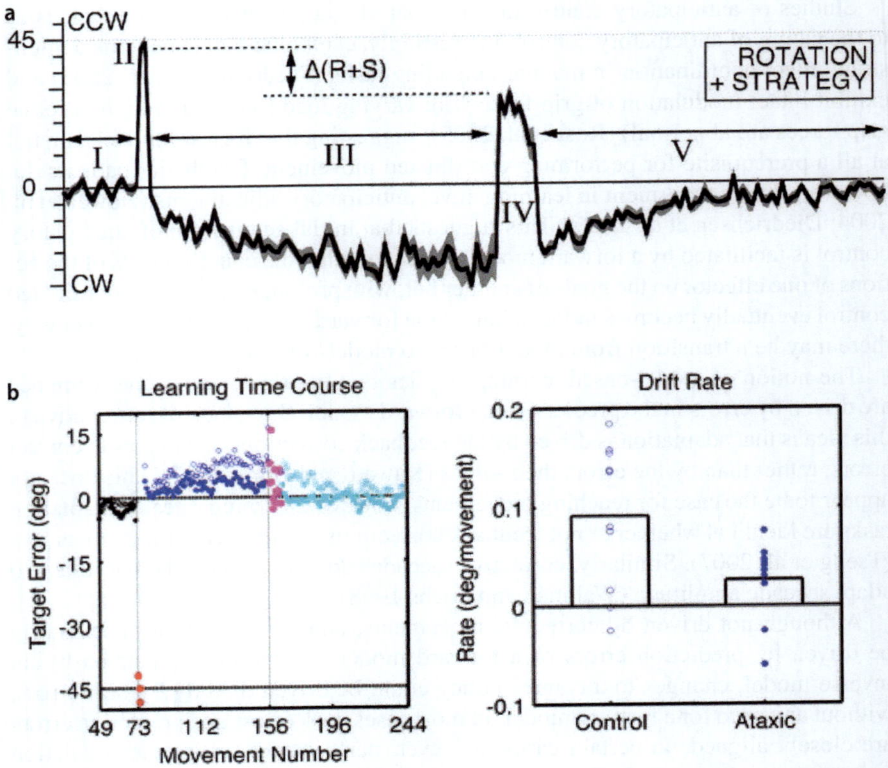

Fig. 1.2 Motor learning is driven by sensory prediction errors. **a** Healthy subjects that are provided with an explicit strategy to counter a 45° rotation initially counter the perturbation successfully, but performance immediately drifts in the direction opposite the rotation. (Reproduced from Mazzoni and Krakauer 2006). **b** This drift is attenuated in patients with cerebellar ataxia (note that rotation direction is opposite as compared with panel **a**). (Reproduced from Taylor et al. 2010)

acts to close task errors rather than prediction errors (Taylor and Ivry 2011). When patients with cerebellar ataxia are given an explicit strategy, they are able to successfully maintain performance without undergoing any drift in performance (Taylor et al. 2010) (Fig. 1.2b). Thus the adaptation deficit in cerebellar ataxic patients is due to a reduced sensitivity to prediction errors not task errors.

The idea that adaptation is mediated by changes in predictions about the consequences of one's actions can be tested more directly through paradigms that ask subjects to estimate where they perceived their hand to have moved during a reach. Although such assays inevitably contaminate forward model-based predictions with actual visual and proprioceptive sensory experiences, a number of interesting results have been obtained using this approach. Following exposure to rotated visual feedback, healthy subjects undergo a corresponding change in their perceived hand path during movement (Synofzik et al. 2006). Cerebellar ataxic patients show no such perceptual changes (Synofzik et al. 2008; Izawa et al. 2011). These results support the idea that changes in a forward model, which presumably lead to the changes in predicted hand position, are a prerequisite for adaptation.

In summary, adaptation is driven by prediction errors and not by task errors or online motor responses to correct those errors. Exposure to rotated visual feedback leads to a shift in perceived hand location during movement. In patients with cerebellar ataxia, sensory prediction errors do not result in changes in feedforward control in future trials and do not lead to changes in perceived hand position. We believe that the most parsimonious explanation for all of these results is that the cerebellum computes an internal forward model that predicts the consequences of motor commands and that this forward model influences feedforward control of future movements.

Generalization of Learning Across Tasks

A final thread of evidence that has been cited in support of model-based control frameworks concerns generalization. Human subjects exhibit a high degree of generalization of learned compensation for a perturbation to a new movement (Shadmehr and Mussa-Ivaldi 1994; Krakauer et al. 2000). While this generalization is consistent with the idea of model-based control, it is important to bear in mind that model-free learning will also be expected to exhibit some degree of generalization—only in this case the generalization will be of a learned control policy, rather than of an internal model. The amount of generalization across states will be entirely determined by the underlying representation. There is no specific reason why one should expect model-based learning to generalize more broadly across states than model-free. However, subjects trained on a visuomotor rotation with full vector error (presumably engaging primarily model-based mechanisms) do generalize more broadly than subjects who learned to compensate the same perturbation but were given only binary feedback about the success or failure of their movements (presumably relying on model-free learning) (Izawa and Shadmehr 2011).

A more concrete dissociation between model-based and model-free learning mechanisms is the extent to which learning should transfer across tasks within the same workspace—for instance tracking a cursor along a curved path versus making point-to-point reaches. This form of generalization across tasks is directly analogous to the reward devaluation protocols that dissociate model-based from model-free action selection processes in rodents (Daw et al. 2005)—in both cases the reward structure of the task is altered but the consequences of actions remain the same. A number of studies have examined generalization of learning from a redundant task, in which the perturbation is task-irrelevant, to a nonredundant task. For example, Schaefer et al. (2012) had subjects make reaching movements to a point anywhere on a circular arc while imposing a rotation of visual feedback. This rotation did not compromise task performance, since subjects still easily landed on the arc as required. The rotation, however, did lead to sensory prediction errors. In subsequent catch trials toward a single target, subjects showed significant aftereffects, supporting the idea that learning was driven by sensory prediction errors rather than by task errors. An identical pattern was found for subjects who adapted to a visual amplification of movement extent (gain increase) while performing an analogous task in which reach direction mattered, but reach extent did not.

Interestingly, although perturbations led to significant aftereffects even when they produced errors only along task-irrelevant dimensions, the aftereffects found by Schaefer and colleagues were significantly smaller than when the same perturbations were task-relevant. A similar study by Synofzik and colleagues (2006) found similar partial transfer of a visual rotation from reaching to an arc to reaching to a point. In that study, additional probe trials measured perceptual changes accompanying learning. Significant changes in perceived hand position during reaching were found. However, although the mean perceptual changes and mean generalization were of similar magnitude, individual subjects' perceptual changes were not predictive of the reaching behavior. Incomplete transfer of adaptation across tasks is not limited to these two examples. Adaptation to a shift in visual feedback during a manual tracking task generalizes only partially to subsequent reaching movements (Simani et al. 2007). In force field adaptation, subjects who learn a force field while performing a series of point-to-point reaching movements show partial but incomplete generalization of their learning when subsequently asked to make circular movements (Conditt et al. 1997).

The finding that learning is consistently seen to transfer across tasks is consistent with model-based learning. However, the variation in the extent of learning under task-relevant and task-irrelevant conditions suggests that learning might not be purely driven by changes in a forward model. On the one hand, one could interpret these results in terms of task-specific internal models mediating task-specific model-based controllers. However, the notion of task-specific internal models rather defeats one of the primary benefits of a model-based approach to control: the ability to flexibly generalize knowledge about the environment across tasks. It perhaps seems more parsimonious to suggest that task specificity of learning arises because of the task specificity of components of learning that are independent of internal models.

Evidence for Model-Free Learning

Model-based learning implies that the motor system learns to compensate for systematic perturbations by first identifying the dynamics of the system being controlled through a forward model (likely in the cerebellum), then somehow translating this knowledge into a control policy in the motor cortex. We have argued that this kind of mechanism can parsimoniously account for a variety of experimental results. However, not all aspects of motor learning are well explained by such a model-based framework. Here we outline the evidence that the motor system also relies on direct, model-free learning of actions.

As we have described above, patients with cerebellar ataxia are severely impaired in compensating for systematic perturbations. According to a model-based interpretation, this inability arises from a primary deficit in the ability to learn or update an

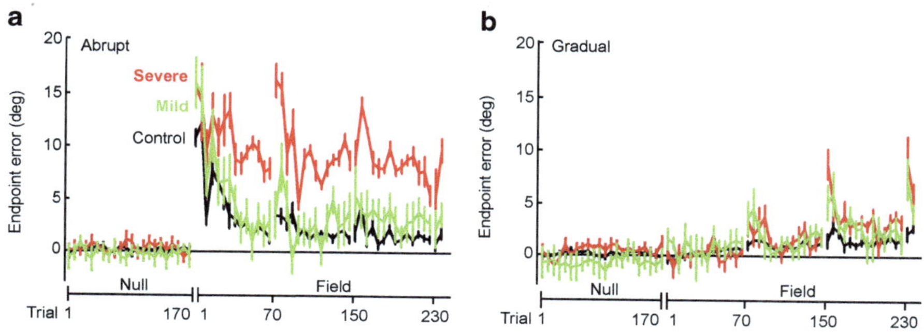

Fig. 1.3 Force field learning in patients with cerebellar ataxia. **a** Severe patients are grossly impaired in adapting to an abruptly introduced force field. **b** Introducing the force field gradually allows compensation similar to healthy controls. (Reproduced from Criscimagna-Hemminger et al. 2010)

internal forward model describing the perturbation (either explicitly, or as an adjustment to baseline forward models). Perhaps surprisingly, however, cerebellar ataxic patients are able to learn to compensate for perturbations if the perturbation is introduced sufficiently gradually (Criscimagna-Hemminger et al. 2010; Izawa et al. 2011) (Fig. 1.3). Successful learning in this case is not accompanied by a change in the perceived consequences of actions (Izawa et al. 2011). This spared learning ability of cerebellar patients therefore does not appear to be associated with any change in an internal model. Interestingly, healthy subjects who learn to compensate for a rotation given only binary feedback about the success or failure of their movements are similarly able to successfully learn to counter the rotation without updating their predictions about the outcome of their movements (Izawa and Shadmehr 2011). These findings suggests that cerebellar ataxia patients, as well as healthy controls deprived of being able to extract prediction errors from a movement, are able to compensate by engaging an alternative learning mechanism that relies solely on the degree of success of a movement and not on the directionality or magnitude of errors. This also explains why learning is only possible in a gradual paradigm: natural variability in movements ensures occasional success when perturbations are small, sufficient to allow the patients to shift their control strategy and in turn enabling them to find movements that successfully counter larger perturbations later on.

The clearest example where model-based and model-free learning can be clearly dissociated is in explaining the ability to relearn a perturbation faster the second time around (savings). Huang and colleagues (2011) had subjects reach to a series of targets, while giving them rotated visual feedback. Crucially, the rotation changed pseudo-randomly from trial to trial, so that subjects achieved little success and never repeatedly made the same movement with success. Subjects did, however, adapt to the average imposed perturbation, which was a clockwise rotation of 20°. Following washout of this learning, subjects were given a further test block in which they made reaching movements to a single target under a constant rotation. In this test block, these subjects adapted no faster than naïve subjects, i.e., they did not exhibit savings despite having previously adapted their movements in the direction appropriate to

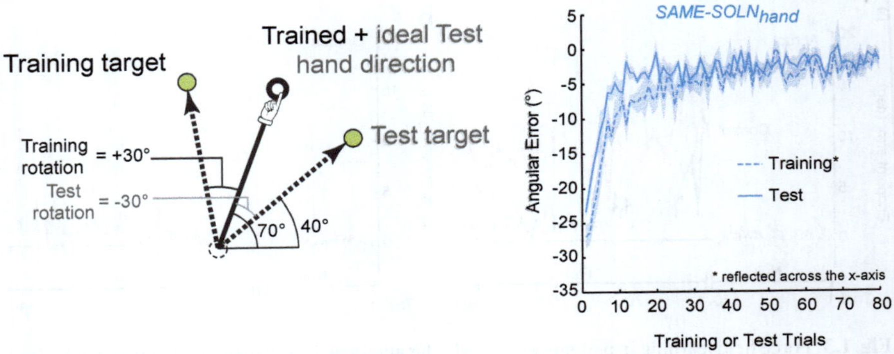

Fig. 1.4 Adaptation to a visual rotation is accelerated by prior learning of an opposite rotation, provided the targets are arranged such that the action required to successfully counter the rotation is the same in both cases. (Reproduced from Huang et al. 2011)

counter the test perturbation. By contrast, a second group of subjects was faced with a nonrandom perturbation that was designed such that the same action would lead to success on every trial, regardless of the target location. After washout, subjects in this group adapted substantially faster to a subsequent test rotation, i.e., they exhibited strong savings. These results suggest that learning an internal model alone does not suffice to achieve savings—some degree of repetition of a successful movement is also necessary. Even so, the savings might have been due to a model-based process that is modulated by task success. A further experiment established that this is not the case by showing that savings can arise from prior learning of an *opposite rotation*, provided that both rotations required the same hand movement to achieve success (Fig. 1.4).

These results suggest that savings arises through recall of previously successful actions, rather than recall of a learned internal model. It may even be that the recall of successful actions is prompted by reward prediction errors rather than by sensory prediction errors. In force field learning, withholding a previously given reinforcement signal triggers partial recall of previously learned actions, even when kinematic performance errors are mechanically clamped at zero (Pekny et al. 2011).

Dopamine-dependence of Model-free Learning

From a theoretical perspective, model-free learning relies on errors in predicted reward (unlike model-based learning, which relies on errors in predicted sensory feedback). Dopamine neurons have been consistently linked with reward prediction errors (Montague et al. 1996; Schultz et al. 1997). Parkinson's disease (PD), in which there is widespread death of dopaminergic neurons, is known to lead to learning deficits in operant learning tasks that rely on reward prediction error signals (Frank et al. 2004; Shohamy et al. 2005; Avila et al. 2009). Thus the study of learning in

PD patients may offer crucial insights into which components of motor learning are subserved by model-free processes. PD patients show no impairment in learning a visuomotor rotation compared with age-matched control subjects. However, savings upon re-adaptation is almost absent in patients with PD (Marinelli et al. 2009; Bedard and Sanes 2011; Leow et al. 2012). This remarkable finding clearly demonstrates the importance of reinforcement and reward for savings and strongly accords with the results of Huang and colleagues (2011). Dopamine is also known to play a pivotal role in skill acquisition in rats. Blocking dopaminergic innervation of M1 from the ventral tegmental area (VTA) abolishes the ability of rats to improve performance in a grasping task. Blocking dopamine did not, however, impair performance of previously acquired skills, suggesting that dopamine plays a key role in learning but not execution of motor skill (Hosp et al. 2011).

Use-Dependent Learning

A significant line of observations that is difficult to reconcile with purely model-based frameworks is the fact that, even in the absence of perturbations, current movements appear to be influenced by previous movements—a phenomenon that has been termed *use-dependent learning*. Repeated movements toward a particular target lead future movements to be biased toward that movement direction (Huang et al. 2011; Verstynen and Sabes 2011). Point-to-point reaching movements around an obstacle lead to a distinct trajectory bias once the obstacle has been removed (Jax and Rosenbaum 2007). In redundant tasks, subjects are biased toward solutions that they were led to on previous attempts, even though these may be far from energetically optimal (Diedrichsen et al. 2010). These history-dependent biases in behavior may originate in primary motor cortex: motor responses elicited by transcranial magnetic stimulation (TMS) over M1 tend to be biased toward movements that were practiced immediately beforehand (Classen et al. 1998). Long-term practice over years seems to have a similar effect—TMS of M1 in expert musicians is more likely to elicit the same hand movements that occur while playing their instrument compared to nonmusician controls, or musicians that play other instruments (Gentner et al. 2010).

Use-dependent learning appears to be difficult to reconcile with the model-based view that movements are planned to be optimal according to a current model of the environment. Use-dependent learning also does not seem to be due to the same model-free learning mechanisms that give rise to savings. Movement execution biases can be induced without giving rise to savings (Huang et al. 2011), and savings occurs when target directions are distributed uniformly around a single start position, in which case no use-dependent learning occurs (Verstynen and Sabes 2011). It can be argued that these kinds of history-dependent effects reflect a learned model of the structure of the task, with biases in movement direction reflecting the influence of prior expectations about the location of the target (Verstynen and Sabes 2011). In this case, use-dependent learning can be thought of as a form of unsupervised learning, being driven not by prediction errors or rewards but by the statistics of previous actions.

An alternative way to frame use-dependent learning is as a form of motor *habit*–i.e. an insensitivity to changes in task goals (which is a generalization of the notion of reward revaluation). Habit and model-free learning have been equated in the realm of rodent decision-making (Daw et al. 2005). It may be possible, however, to dissociate them in the domain of motor control where, critically, independent assays of habit (movement biases) and model-free learning (savings) are available.

In summary, there are an increasing number of experimental observations in motor learning that are difficult to describe within model-based frameworks. We do not wish to suggest that these findings negate the idea of model-based learning. Rather, we propose that these phenomena occur due to additional learning mechanisms that operate independently of internal models. Furthermore, these mechanisms constitute more than just a curious nuisance that contaminates behavior in adaptation paradigms. We suggest that they are equally if not more critical than model-based learning. Indeed, acquisition of entirely novel motor behaviors may depend upon model-free learning mechanisms in which successful control policies in motor cortex are dopaminergically reinforced.

Combining Model-Based and Model-Free Learning Mechanisms

In sequential decision-making contexts, model-based and model-free learning systems are often conceived as operating independently, interacting only at the action selection stage. Both model-based and model-free strategies yield estimates of the value of choosing a particular action in a particular state, and these independent estimates must be combined to guide the ultimate choice of action. The confidence in each estimate plays a crucial role in this arbitration process—a concept that can formalized in a Bayesian sense (Daw et al. 2005). The ultimate course of action could be based on either choosing the most reliable estimate (Daw et al. 2005) or combing the estimated values weighted by their relative reliabilities (Glascher et al. 2010). Early in learning, when there are few observations available, model-based approaches tend to be more reliable than model-free ones. Imprecise computations place a limit on the reliability of model-based methods, however, so that, when data are abundant, the direct approach of model-free learning becomes more reliable. Consequently, behavior tends to rely more on model-based mechanisms early on and more on model-free mechanisms after extensive practice.

In the context of control of the eyes and limbs, there is a continuous, possibly high-dimensional space of potential actions. When choosing among a discrete set of actions, an animal can exhaustively sample all available actions in each state—the only constraints on learning are the confidence that the animal has in its own observations and in the constancy of the environment. When the action space is continuous-valued, however, exhaustive sampling is not possible. In these scenarios, the model-based system becomes essential to rapidly guide behavior toward promising control policies, effectively guiding exploration for model-free learning. As is the case in discrete domains, abilities that are initially learned model-based should, with experience, become model-free behaviors. This transfer of responsibility explains the fact that anticipatory control remains intact in patients with cerebellar ataxia (who will have initially learned such coordination long before the onset of ataxia (Diedrichsen et al. 2005)) but new patterns of coordination cannot be learned.

Conclusions and Outlook

We have contrasted two distinct approaches to learning new control policies: model-based learning, in which sensory prediction errors indirectly drive updates to a control policy by updating a forward model, and model-free learning in which reward prediction errors drive changes to a control policy directly. We have argued that the motor system utilizes both kinds of learning. A parallel learning architecture lends the motor system robustness and redundancy; learning a given task can be achieved in many different ways, so that if one fails either through circumstance (such as if impoverished sensory feedback precludes learning a forward model) or disease, other mechanisms are still in place to ensure that overall performance can still be maintained or improved.

It has long been argued that the different structures within the brain appear to be well suited to learning from experience in different ways (Doya 1999).The complementary roles of different brain regions for model-based and model-free learning are particularly nicely exhibited in a recent study by Galea and colleagues (2011). They found that anodal transcranial direct current stimulation (tDCS) over primary motor cortex did not affect the rate of adaptation to a visuomotor rotation. However, it did lead to a large effect on retention of learning. Conversely, anodal tDCS of the cerebellum substantially increased the rate of adaptation, but did not influence retention. Although the exact mechanisms by which brain stimulation is able to modulate learning in this way are far from clear at present, we suggest a general way of viewing this result as tDCS of the cerebellum promoting model-based learning and tDCS of M1 promoting model-free learning.

It is likely that the brain utilizes many different forms of learning, not restricted to the specific strategies and mechanisms we have focused on here. For instance, although we have described model-based learning as being dependent on an implicit internal model in the cerebellum, it is likely that other brain areas such as prefrontal cortex may contribute alternative, explicit task models that may give rise to a form of model-based control that might be considered more strategic (Mazzoni and Krakauer 2006; Taylor and Ivry 2011). Use-dependent learning appears to be neither model-based nor model-free in the sense that we have described. Many error-driven learning strategies have been proposed in which vector performance errors directly drive updates to a controller (Thoroughman and Shadmehr 2000; Franklin et al. 2008). Such strategies are neither model-based in the sense we have described (since the error directly influences the controller, rather than going through a forward model) nor model-free (since learning is error-driven rather than reward-driven).

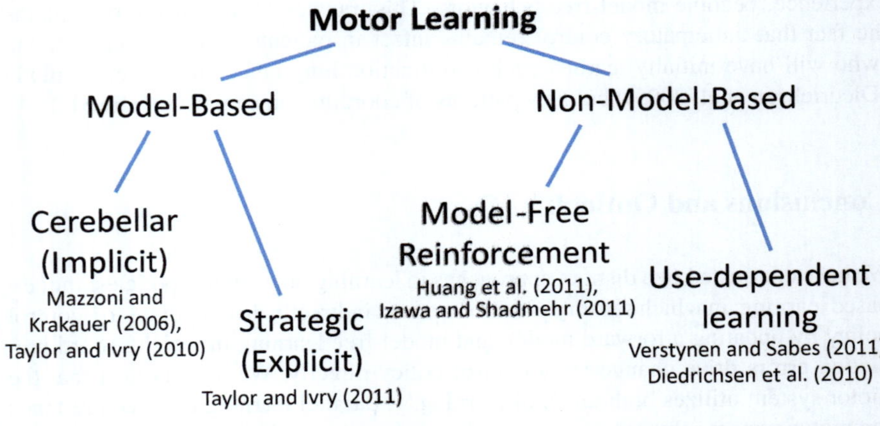

Fig. 1.5 A motor learning taxonomy, by analogy to the famous taxonomy of memory by Squire. (Squire 1992)

A deeper understanding of the multiple component processes that support motor learning is paramount to advance the efficacy of neuro-rehabilitation following brain injury. The most promising path to achieving this goal is through coupling theoretical insights with carefully designed experiments, study of specific patient groups, and the use of brain stimulation techniques. Dissecting motor learning into its constituent components is a clear and important goal for future motor control research. We believe that computational levels of description offer a sound basis by which to begin this classification (Fig. 1.5).

References

Ariff G, Donchin O, Nanayakkara T, Shadmehr R (2002) A real-time state predictor in motor control: study of saccadic eye movements during unseen reaching movements. J Neurosci 22:7721–7729

Avila I, Reilly MP, Sanabria F, Posadas-Sanchez D, Chavez CL, Banerjee N, Killeen P, Castaneda E (2009) Modeling operant behavior in the Parkinsonian rat. Behav Brain Res 198:298–305

Balleine BW, Dickinson A (1998) Goal-directed instrumental action: contingency and incentive learning and their cortical substrates. Neuropharmacology 37:407–419

Balleine BW, O'Doherty JP (2010) Human and rodent homologies in action control: corticostriatal determinants of goal-directed and habitual action. Neuropsychopharmacology 35:48–69

Bastian AJ (2006) Learning to predict the future: the cerebellum adapts feedforward movement control. Curr Opin Neurobiol 16:645–649

Bedard P, Sanes JN (2011) Basal ganglia-dependent processes in recalling learned visual-motor adaptations. Exp Brain Res 209:385–393

Bertsekas DP (1996) Dynamic programming and optimal control: Athena Scientific

Carmena JM, Lebedev MA, Crist RE, O'Doherty JE, Santucci DM, Dimitrov DF, Patil PG, Henriquez CS, Nicolelis MA (2003) Learning to control a brain-machine interface for reaching and grasping by primates. PLoS Biol 1:E42

Chen-Harris H, Joiner WM, Ethier V, Zee DS, Shadmehr R (2008) Adaptive control of saccades via internal feedback. J Neurosci 28:2804–2813

Classen J, Liepert J, Wise SP, Hallett M, Cohen LG (1998) Rapid plasticity of human cortical movement representation induced by practice. J Neurophysiol 79:1117–1123

Conditt MA, Gandolfo F, Mussa-Ivaldi FA (1997) The motor system does not learn the dynamics of the arm by rote memorization of past experience. J Neurophysiol 78:554–560

Criscimagna-Hemminger SE, Bastian AJ, Shadmehr R (2010) Size of error affects cerebellar contributions to motor learning. J Neurophysiol 103:2275–2284

Daw ND, Niv Y, Dayan P (2005) Uncertainty-based competition between prefrontal and dorsolateral striatal systems for behavioral control. Nat Neurosci 8:1704–1711

Dayan P (2009) Goal-directed control and its antipodes. Neural Netw 22:213–219

Diedrichsen J, Verstynen T, Lehman SL, Ivry RB (2005) Cerebellar involvement in anticipating the consequences of self-produced actions during bimanual movements. J Neurophysiol 93:801–812

Diedrichsen J, White O, Newman D, Lally N (2010) Use-dependent and error-based learning of motor behaviors. J Neurosci 30:5159–5166

Donchin O, Francis JT, Shadmehr R (2003) Quantifying generalization from trial-by-trial behavior of adaptive systems that learn with basis functions: theory and experiments in human motor control. J Neurosci 23:9032–9045

Donchin O, Rabe K, Diedrichsen J, Lally N, Schoch B, Gizewski ER, Timmann D (2011) Cerebellar regions involved in adaptation to force field and visuomotor perturbation. J Neurophysiol 107(1):134–47

Doya K (1999) What are the computations of the cerebellum, the basal ganglia and the cerebral cortex? Neural Netw 12:961–974

Fermin A, Yoshida T, Ito M, Yoshimoto J, Doya K (2010) Evidence for model-based action planning in a sequential finger movement task. J Mot Behav 42:371–379

Flanagan JR, Wing AM (1997) The role of internal models in motion planning and control: evidence from grip force adjustments during movements of hand-held loads. J Neurosci 17:1519–1528

Frank MJ, Seeberger LC, O'Reilly R C (2004) By carrot or by stick: cognitive reinforcement learning in parkinsonism. Science 306:1940–1943

Franklin DW, Burdet E, Tee KP, Osu R, Chew CM, Milner TE, Kawato M (2008) CNS learns stable, accurate, and efficient movements using a simple algorithm. J Neurosci 28:11165–11173

Galea JM, Vazquez A, Pasricha N, de Xivry JJ, Celnik P (2011) Dissociating the roles of the cerebellum and motor cortex during adaptive learning: the motor cortex retains what the cerebellum learns. Cereb Cortex 21:1761–1770

Gentner R, Gorges S, Weise D, aufm Kampe K, Buttmann M, Classen J (2010) Encoding of motor skill in the corticomuscular system of musicians. Curr Biol 20:1869–1874

Glascher J, Daw N, Dayan P, O'Doherty JP (2010) States versus rewards: dissociable neural prediction error signals underlying model-based and model-free reinforcement learning. Neuron 66:585–595

Hosp JA, Pekanovic A, Rioult-Pedotti MS, Luft AR (2011) Dopaminergic projections from midbrain to primary motor cortex mediate motor skill learning. J Neurosci 31:2481–2487

Huang VS, Haith A, Mazzoni P, Krakauer JW (2011) Rethinking motor learning and savings in adaptation paradigms: model-free memory for successful actions combines with internal models. Neuron 70:787–801

Izawa J, Shadmehr R (2011) Learning from sensory and reward prediction errors during motor adaptation. PLoS Comput Biol 7:e1002012

Izawa J, Criscimagna-Hemminger SE, Shadmehr R (2011) Cerebellar Contributions to Learning Sensory Consequences of Action. J Neurosci 32(12):4230–4239

Jax SA, Rosenbaum DA (2007) Hand path priming in manual obstacle avoidance: evidence that the dorsal stream does not only control visually guided actions in real time. J Exp Psychol 33:425–441

Jordan MIaR, D.E. (1992) Forward models: Supervised learning with a distal teacher. Cognitive Sci 16:307–354

Kawato M, Gomi H (1992) A computational model of four regions of the cerebellum based on feedback-error learning. Biol Cybern 68:95–103

Killcross S, Coutureau E (2003) Coordination of actions and habits in the medial prefrontal cortex of rats. Cereb Cortex 13:400–408
Krakauer JW, Ghilardi MF, Ghez C (1999) Independent learning of internal models for kinematic and dynamic control of reaching. Nat Neurosci 2:1026–1031
Krakauer JW, Pine ZM, Ghilardi MF, Ghez C (2000) Learning of visuomotor transformations for vectorial planning of reaching trajectories. J Neurosci 20:8916–8924
Lackner JR, Dizio P (1994) Rapid adaptation to Coriolis force perturbations of arm trajectory. J Neurophysiol 72:299–313
Leow LA, Loftus AM and Hammond GR (2012) Impaired savings despite intact initial learning of motor adaptation in Parkinson's disease. Exp Brain Res 2:295–304
Marinelli L, Crupi D, Di Rocco A, Bove M, Eidelberg D, Abbruzzese G, Ghilardi MF (2009) Learning and consolidation of visuo-motor adaptation in Parkinson's disease. Parkinsonism Relat Disord 15:6–11
Martin TA, Keating JG, Goodkin HP, Bastian AJ, Thach WT (1996) Throwing while looking through prisms. I. Focal olivocerebellar lesions impair adaptation. Brain 119 (4):1183–1198
Maschke M, Gomez CM, Ebner TJ, Konczak J (2004) Hereditary cerebellar ataxia progressively impairs force adaptation during goal-directed arm movements. J Neurophysiol 91:230–238
Mazzoni P, Krakauer JW (2006) An implicit plan overrides an explicit strategy during visuomotor adaptation. J Neurosci 26:3642–3645
McGuire LM, Sabes PN (2009) Sensory transformations and the use of multiple reference frames for reach planning. Nat Neurosci 12:1056–1061
Medina JF (2011) The multiple roles of Purkinje cells in sensori-motor calibration: to predict, teach and command. Curr Opin Neurobiol 21:616–622
Mehta B, Schaal S (2002) Forward models in visuomotor control. J Neurophysiol 88:942–953
Miall RC, Christensen LO, Cain O, Stanley J (2007) Disruption of state estimation in the human lateral cerebellum. PLoS Biol 5:e316
Montague PR, Dayan P, Sejnowski TJ (1996) A framework for mesencephalic dopamine systems based on predictive Hebbian learning. J Neurosci 16:1936–1947
Mosier KM, Scheidt RA, Acosta S, Mussa-Ivaldi FA (2005) Remapping hand movements in a novel geometrical environment. J Neurophysiol 94:4362–4372
Munuera J, Morel P, Duhamel JR, Deneve S (2009) Optimal sensorimotor control in eye movement sequences. J Neurosci 29:3026–3035
Nagengast AJ, Braun DA, Wolpert DM (2009) Optimal control predicts human performance on objects with internal degrees of freedom. PLoS Comput Biol 5:e1000419
Nowak DA, Hermsdorfer J, Rost K, Timmann D, Topka H (2004) Predictive and reactive finger force control during catching in cerebellar degeneration. Cerebellum 3:227–235
Pasalar S, Roitman AV, Durfee WK, Ebner TJ (2006) Force field effects on cerebellar Purkinje cell discharge with implications for internal models. Nat Neurosci 9:1404–1411
Pekny SE, Criscimagna-Hemminger SE, Shadmehr R (2011) Protection and expression of human motor memories. J Neurosci 31:13829–13839
Rabe K, Livne O, Gizewski ER, Aurich V, Beck A, Timmann D, Donchin O (2009) Adaptation to visuomotor rotation and force field perturbation is correlated to different brain areas in patients with cerebellar degeneration. J Neurophysiol 101:1961–1971
Reis J, Schambra HM, Cohen LG, Buch ER, Fritsch B, Zarahn E, Celnik PA, Krakauer JW (2009) Noninvasive cortical stimulation enhances motor skill acquisition over multiple days through an effect on consolidation. Proc Natl Acad Sci 106:1590–1595
Roitman AV, Pasalar S, Johnson MT, Ebner TJ (2005) Position, direction of movement, and speed tuning of cerebellar Purkinje cells during circular manual tracking in monkey. J Neurosci 25:9244–9257
Rost K, Nowak DA, Timmann D, Hermsdorfer J (2005) Preserved and impaired aspects of predictive grip force control in cerebellar patients. Clin Neurophysiol 116:1405–1414
Schaefer SY, Shelly IL, Thoroughman KA (2012) Beside the point: motor adaptation without feedback-based error correction in task-irrelevant conditions. J Neurophysiol 107(4):1247–56
Schultz W, Dayan P, Montague PR (1997) A neural substrate of prediction and reward. Science 275:1593–1599.

Shadmehr R, Mussa-Ivaldi FA (1994) Adaptive representation of dynamics during learning of a motor task. J Neurosci 14:3208–3224

Shadmehr R, Krakauer JW (2008) A computational neuroanatomy for motor control. Exp Brain Res 185:359–381

Shadmehr R, Smith MA, Krakauer JW (2010) Error correction, sensory prediction, and adaptation in motor control. Annu Rev Neurosci 33:89–108

Shmuelof L, Krakauer JW, Mazzoni P (2012) How is a motor skill learned? Change and invariance at the levels of task success and trajectory control. J Neurophysiol 08(2):578–94

Shohamy D, Myers CE, Grossman S, Sage J, Gluck MA (2005) The role of dopamine in cognitive sequence learning: evidence from Parkinson's disease. Behav Brain Res 156:191–199

Simani MC, McGuire LM, Sabes PN (2007) Visual-shift adaptation is composed of separable sensory and task-dependent effects. J Neurophysiol 98:2827–2841

Smith MA, Shadmehr R (2005) Intact ability to learn internal models of arm dynamics in Huntington's disease but not cerebellar degeneration. J Neurophysiol 93:2809–2821

Squire LR (1992) Memory and the hippocampus: a synthesis of findings with rats, monkeys and humans. Psychol Rev 99:195–231

Sternad D, Abe MO, Hu X, Muller H (2011) Neuromotor noise, error tolerance and velocity-dependent costs in skilled performance. PLoS Comput Biol 7:e1002159

Sutton RS, Barto AG (1998) Reinforcement learning: An introduction. Cambridge Univ Press

Synofzik M, Thier P, Lindner A (2006) Internalizing agency of self-action: perception of one's own hand movements depends on an adaptable prediction about the sensory action outcome. J Neurophysiol 96:1592–1601

Synofzik M, Lindner A, Thier P (2008) The cerebellum updates predictions about the visual consequences of one's behavior. Curr Biol 18:814–818

Taylor JA, Ivry RB (2011) Flexible cognitive strategies during motor learning. PLoS Comput Biol 7:e1001096

Taylor JA, Klemfuss NM, Ivry RB (2010) An explicit strategy prevails when the cerebellum fails to compute movement errors. Cerebellum 9:580–586

Thoroughman KA, Shadmehr R (2000) Learning of action through adaptive combination of motor primitives. Nature 407:742–747

Todorov E (2007) Optimal control theory. In: Bayesian brain: probabilistic approaches to neural coding, MIT Press, Cambridge, p 269–298

Todorov E, Jordan MI (2002) Optimal feedback control as a theory of motor coordination. Nat Neurosci 5:1226–1235

Tseng YW, Diedrichsen J, Krakauer JW, Shadmehr R, Bastian AJ (2007) Sensory prediction errors drive cerebellum-dependent adaptation of reaching. J Neurophysiol 98:54–62

van der Meer MA, Redish AD (2011) Ventral striatum: a critical look at models of learning and evaluation. Curr Opin Neurobiol 21:387–392

Verstynen T, Sabes PN (2011) How each movement changes the next: an experimental and theoretical study of fast adaptive priors in reaching. J Neurosci 31:10050–10059

Wallman J, Fuchs AF (1998) Saccadic gain modification: visual error drives motor adaptation. J Neurophysiol 80:2405–2416

Wagner MJ, Smith MA (2008) Shared internal models for feedforward and feedback control. J Neurosci 28:10663–10673

Wolpert DM, Miall RC (1996) Forward Models for Physiological Motor Control. Neural Netw 9:1265–1279

Wolpert DM, Miall RC, Kawato M (1998) Internal models in the cerebellum. Trends Cogn Sci 2:338–347

Wolpert DM, Diedrichsen J, Flanagan JR (2011) Principles of sensorimotor learning. Nat Rev Neurosci 12:739–751

Wong AL, Shelhamer M (2011) Sensorimotor adaptation signals are derived from realistic predictions of movement outcomes. J Neurophysiol 105(3):1130–40

Xu-Wilson M, Chen-Harris H, Zee DS, Shadmehr R (2009) Cerebellar contributions to adaptive control of saccades in humans. J Neurosci 29:12930–12939

Chapter 2
The Molecular Basis of Experience-Dependent Motor System Development

Robert G. Kalb, Weiguo Zhou and Lei Zhang

Overview

Nervous system operation depends on the ability of neurons to process a panoply of excitatory and inhibitory signals and generate meaningful output. The size and complexity of the dendritic tree is a critical determinant of the computational work of neurons (Stuart et al. 1999). In addition, the geometry of the dendritic tree can regulate who communicates with a neuron by controlling the quantitative and qualitative nature of the afferent input (Hume and Purves 1981). Theorists, taking a wiring optimization approach, view synaptic connectivity and neuronal morphology as inextricably linked because it is the most efficient fit of network wiring within a given volume of neuropil (Chklovskii 2004). The spectrum of animal behavior, from a 0.5 mm roundworm wiggling on a Petri dish to Glen Gould playing Bach, reflects the precision with which neurons elaborate their dendritic tree and are innervated.

R. G. Kalb
Department of Neurology, Children's Hospital of Philadelphia, Perelman School of Medicine, University of Pennsylvania, 3615 Civic Center Boulevard, Philadelphia, PA 19104, USA
e-mail: kalb@email.chop.edu

R. G. Kalb · W. Zhou · L. Zhang
Research Institute and Division of Neurology, Department of Pediatrics, Children's Hospital of Philadelphia, Perelman School of Medicine, University of Pennsylvania, 3615 Civic Center Boulevard, Philadelphia, PA 19104, USA

W. Zhou
e-mail: weiguozhou@gmail.com

L. Zhang
e-mail: ZhangL@email.chop.edu

M. J. Richardson et al. (eds.), *Progress in Motor Control,*
Advances in Experimental Medicine and Biology 782,
DOI 10.1007/978-1-4614-5465-6_2,

Cell and Molecular Biology of Activity-Dependent Development

The process of dendrite elaboration is often divided into an initial, synaptic activity-independent phase and a subsequent, synaptic activity-dependent phase (Goodman and Shatz 1993; Shatz 1990). The first phase sets up the basic architecture of the tree and is likely to be under strong genetic control (Gao and Bogert 2003; Jan and Jan 2003). The synaptic activity-dependent phase of dendrite elaboration is believed to fine tune structure into a precise configuration (Cline 2001; Constantine-Paton et al. 1990). There is much evidence to support the "synaptotrophic hypothesis" as a mechanism for activity-dependent regulation of dendrite arbor development (Vaughn 1989; Vaughn et al. 1988; reviewed by Cline and Haas 2008). Developing axons and dendrites undergo exploratory growth and make nascent synapses. Adhesion molecules such as neurexins (NRX) and neuroligins (NLG; Chen et al. 2010; Thyagarajan and Ting 2010) are likely to be involved. Repeated use of a synapse leads to stabilization of the synapse if pre- and postsynaptic elements are coincidently active (Bi and Poo 2001; Engert et al. 2002; Ruthazer et al. 2003). Axons and dendrites that bear stable synapses are retained (or perhaps grow) and conversely, portions of axons and dendrites that do not bear stable synapses are withdrawn (Katz and Constantine-Paton 1988). Spontaneous synaptic activity is the initial driver of these events and subsequently environmentally evoked synaptic activity does the heavy lifting. Experience-dependent refinement of neuronal architecture and synaptic connectivity sculpts each nervous system to perform best in the environment in which the animal was reared (Zhang et al. 2000).

One form of activity-dependent development involves activation of the N-methyl-D-aspartate (NMDA) subtype of glutamate receptor (Cline et al. 1987; Kalb 1994; Kleinschmidt et al. 1987). This leads to a substantial rise in dendritic calcium, which is thought to be the key trigger for subsequent events. The precise ordering of what occurs next is not entirely clear, but the literature supports the view that there are three linked main events. Event #1 is the secretion/elaboration of extracellular factors such as BDNF, Wnts, and nitric oxide (Cramer et al. 1996; Inglis et al. 1998; Wu et al. 1994). There is evidence that secretion/elaboration is activity-dependent and blocking their action can prevent synapse stablization and dendrite growth (McAllister et al. 1996, 1997; also see Lu 2003). Event #2 is the activation of intracellular signaling molecules such as Ca^{++}/calmodulin-dependent kinase (CamK) type I (Wayman et al. 2006), CamK II (Wu and Cline 1998; Zou and Cline 1999; Gaudilliere et al. 2004) and CamK IV (Redmond et al. 2002), mitogen-activated protein (MAP) kinase (Ha and Redmond 2008; Redmond et al. 2002; Wu et al. 2001), βcatenin (Yu and Malenka 2003; Peng et al. 2009), and RhoA GTPases (Li et al. 2000, 2002; Sin et al. 2002). The role of these molecules has been studied using pharmacological inhibitors and expression of dominant-negative and constitutively-active forms of these proteins. Event #3 is a new gene expression and the list of contributing transcription factors includes cAMP response element-binding protein (CREB; Li et al. 2009; Redmond et al. 2002; Wayman et al. 2006), CREST (Aizawa et al. 2004), NeuroD (Gaudilliere et al. 2004), and myocyte-specific enhancer factor

2A (MEF2A; Shalizi et al. 2006). As discussed above, the role of these molecules has been studied employing molecular genetic techniques. Modification of cytoskeletal elements, maturation of silent synapses (NMDA-R only → AMPA-R + NMDA-R), and the precise apposition of pre- and postsynaptic membranes incorporating adhesion molecules are all necessary steps in this process. With so many events apparently occurring simultaneously, it is difficult to discern the epistatic relationships. How activity-dependent processes dovetail with dendrite growth-promoting processes not-shown-to-be-activity-dependent such as activation of PI3K and mammalian target of rapamycin (mTOR; Jaworski et al. 2005; Kumar et al. 2005) and Notch signaling (Redmond et al. 2000) only complicates matters more.

Experience-Dependent Motor System Development

The normal development of the locomotor system (from behavior, to connectivity within the segmental spinal cord, to motor neuron dendrite architecture) emerges during prenatal and early postnatal life (Altman and Sudarshan 1975; Curfs et al. 1993, 1994; Donatelle 1977; Pellis et al. 1991; Seebach and Ziskind-Conhaim 1994; Snider et al. 1992). In the next section, we outline the evidence that locomotor development is experience-dependent and that the molecular machinery that drives this process can involve activation of NMDA-Rs. In addition, we will provide evidence for a second set of molecules that appear to act in parallel with NMDA-Rs to drive motor system development. We have found that GluA1 subunit of the 2-amino-3-(5-methyl-3-oxo-1,2-oxazol-4-yl) propanoic acid (AMPA)-R, in concert with an intracellular-binding partner called SAP97, promote motor system development by an NMDA-R-independent mechanism.

Experience-Dependent Motor System Development and the NMDA-R

We begin with studies by Kerry Walton's group of neonatal rats reared in space. The force of gravity at the surface of the earth is called "1G" and anything less than that is referred to as "microgravity". Walton's group studied a cohort of animals that spent about 2 weeks of early postnatal life in the space shuttle. She showed that young rats that develop in microgravity have demonstrably different locomotor behavior than those that develop on earth (Walton et al. 2005). These observations echo her previous work using the tail suspension model (Walton et al. 1992). Work from my laboratory using these mice demonstrated that the motor neuron dendritic tree also undergoes experience-dependent development (Inglis et al. 2000). The parsimonious construct is that at least some on the alterations in motor neuron dendrite structure subserve the alterations in locomotor function that follow microgravity rearing.

These behavioral and anatomical studies prompted us to ask whether NMDA-Rs were involved in activity-dependent maturation of motor neuron dendritic architecture. We began by asking whether NMDA-R components were expressed by developing motor neurons. In situ hybridization studies show that newborn motor neurons express NR1, NR2A, and NR2C at particularly high levels and over the subsequent next few weeks of life, the abundance of these messenger RNAs (mRNAs) falls off considerably (Stegenga and Kalb 2001). The NR1 subunit undergoes alternative splicing, and an analysis of specific NR1 variants reveals that NR1A, NR1B, NR1-2, and NR1-4 are expressed at particularly high levels in newborn motor neurons. The abundance of these splice variants falls subsequently in early postnatal life (Stegenga and Kalb 2001). This work demonstrates that motor neurons express a unique repertoire of NMDA-R subunits in early postnatal life.

Coincident with the period when motor neurons express a distinct type of NMDA-R, the dendrites of motor neurons are undergoing substantial growth (Curfs et al. 1993; Lindsay et al. 1991; Núñez-Abades et al. 1994). Overall tree size and number of branches increase approximately twofold to threefold between postnatal day 7 (P7) and P21. Antagonism of NMDA-Rs with (2*R*)-amino-5-phosphonopentanoate (APV) or MK-801 inhibits the growth of motor neuron dendrites of NMDA-R (Kalb 1994). In contrast to their effects on developing dendrites, antagonism of NMDA-Rs in adult animals has no effect on motor neuron dendrite architecture. These results indicate that during a critical period in early postnatal life, activation of NMDA-Rs promotes the elaboration of motor neurons dendrites. In subsequent work, we showed that the dendrite growth-promoting actions of NMDA-Rs are mediated by the second messenger, nitric oxide (Inglis et al. 1998). Overall, this work highlights the distinct parallel between the experience-dependent development of sensory and motor systems.

A Novel Form of Activity-Dependent Development Utilizes GluA1-Containing AMPA-R

In addition to characterizing the expression pattern of NMDA-R subunits, we also examined the expression pattern of AMPA receptors (AMPA-R) subunits. Although all subunits undergo developmentally regulated expression, we were impressed that neonatal motor neurons express particularly high levels of the GluA1 (mRNA and protein; Jakowec et al. 1995a, b). (This is the unedited version of the protein that contains the "flip" alternatively spliced exon and unless otherwise stated, we will use "GluA1" to denote "GluA1(Q)flip.") Electrophysiologic studies show that neonatal motor neurons display Ca^{++}-permeable AMPA receptors (as would be expected from assembled tetramers enriched with GluA1 or even homomeric GluA1 tetramers; Carriedo et al. 1996; Vandenberghe et al. 2000). This raises the possibility that the special type of AMPA receptors expressed by neonatal motor neurons is part of the molecular mechanism of experience-dependent dendrite development. The first good clue that this could be the case was a study in which we overexpressed (OE) GluA1 in mature motor neurons (after the period of developmental dendrite growth; Inglis et al. 2002). We found that this led to large-scale remodeling of the dendritic tree with a marked increase in dendrite branching. Expression of a version of GluR1 with an arginine in the critical "Q/R editing site" (GluA1(R)) had no effects on dendrite architecture. AMPA-R assembled from GluA1(R) are calcium impermeable and pass very little current upon activation with glutamate. These in vivo observations suggest that GluA1 can promote dendrite growth and this depends on the ability of GluA1-containing AMPA-R to depolarize cells. Subsequent in vivo and in vitro works provide strong support for the idea that calcium permeability of GluA1-containing AMPA-R is a major determinant of its effect on dendrite growth (Jeong et al. 2006).

One could imagine at least two ways in which increasing the expression of AMPA-R assembled with GluA1 might promote dendrite growth: (1) By enhancing the ability of cells to depolarize upon afferent stimulation, GluA1 might facilitate the activation of NMDA-Rs. In this scenario, GluA1 acts upstream of NMDA-R-mediated events; (2) GluA1 may act through an NMDA-R-independent pathway to regulated dendrite growth. In this scenario, GluA1 acts in parallel to NMDA-R-mediated events. There are several reasons for favoring the second scenario. First, administration of MK-801 to rats OE GluA1 did not block the prodendrite growth effect. We know MK-801 was at an effective concentration in the brain because we could not evoke LTP in the dentate gyrus from animals treated with MK-801 (see Fig. 6 of Inglis et al. 2002). So, OE GluA1 led to increase dendrite branching even though NMDA-Rs were effectively antagonized. Second, NMDA-R mRNAs are developmentally regulated and not expressed by mature motor neurons (Stegenga and Kalb 2001). So, GluA1-mediated dendrite remodeling can occur in neurons that do not express NMDA-R subunits. Finally, we more formally examined the role of NMDA-Rs in GluA1-dependent dendrite growth in vitro.

We grew dissociated spinal cord neurons in vitro and expressed GluA1 by transfection and treated some cells with the NMDA-R open-channel blocker MK-801 (10 μM). Three groups of neurons were studied morphologically: (1) green fluorescent protein (GFP), (2) GFP + GluA1 + vehicle, and (3) GFP + GluA1 + MK-801. MK-801 or vehicle was administered daily to the cultures and after 5 days in vitro (DIV), the cultures were fixed, immunostained for GFP to enhance the cell labeling, camera lucida drawings generated and quantitatively analyzed (Fig. 2.1). Compared with GFP alone, GluA1 led to an ~30% increase in dendrite branches ($F_{(2,60)} = 3.507, p = 0.03$), ~60% increase in overall arbor size ($F_{(2,60)} = 17.005, p < 0.0001$), ~60% increase in average dendrite length ($F_{(2,60)} = 9.370, p = 0.0003$), and a 40 % increase in the length of the longest dendrite ($F_{(2,60)} = 10.817, p < 0.0001$). Treatment with MK-801 did not influence the progrowth effects of GluA1; the dendritic arbor of neurons in the "GluA1 + MK-801" were not statistically different from the dendrite arbor of the neurons in the "GluA1" group. These observations establish that overexpression of GluA1 stimulates dendrite growth in a manner that is independent of NMDA-Rs.

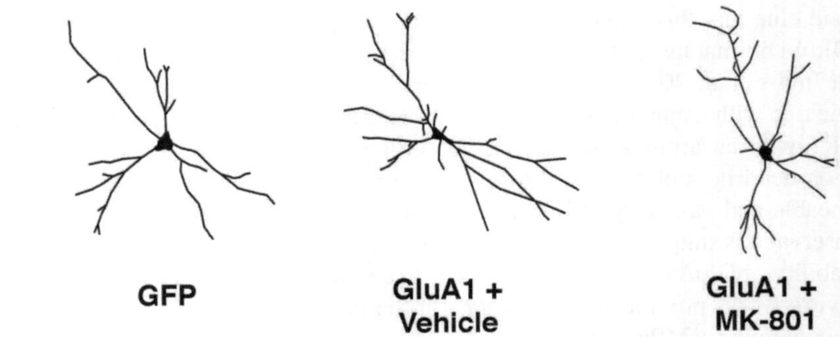

Cell Parameter	GFP Control (n = 20)	GluA1 (n = 21)	MK-801 and GluA1 (n = 22)	$F_{(2,60)}$, *p*
Soma area (μm^2)	308.7 ± 89.9	349.8 ± 77.5	359.2 ± 80.1	2.187; 0.1211
1^{0} dendrites	6.7 ± 1.3	6.8 ± 1.4	6.9 ± 1.9	0.098; 0.9071
Branches, #	13.3 ± 5.2	17.0 ± 4.8*	16.1 ± 4.3	3.507; 0.0363
Σlength (μm)	998.8 ± 236.0	1611.5 ± 431.6***	1581.0 ± 426.5***	17.005; <0.0001
Average dendrites	156.1 ± 46.6	246.3 ± 78.4***	246.6 ± 95.6***	9.370; 0.0003
Longest dendrite (μm)	178.9 ± 57.0	251.9 ± 50.9***	237.3 ± 51.4***	10.817; <0.0001

***Differs from GFP Control: *p<0.05; ** p<0.01; *** p<0.001;**
†Differs from GluR1, †p<0.05; †† p<0.01; ††† p<0.001

Fig. 2.1 Overexpression of wild-type (WT) GluA1 stimulates dendrite growth in vitro in an NMDA-R-independent manner. *Top*, representative camera lucida images of neurons expression GFP alone, or GFP + GluA1 (treated or not with MK-801). The chart *below* provides a quantitative analysis of dendrites as well as a statistical analysis using ANOVA. The number of neurons drawn is noted in parentheses next to the column title. There is statistically significant increase in branching and overall tree size in neurons overexpressing GluA1 and these effects are not influenced by NMDA-R antagonism

In addition to NMDA-R, voltage-gated calcium channels are activated by membrane depolarization, and so these channels might participate in the GluA1 form of dendrite growth. To examine this issue we undertook a second set of experiments; we used the L-type calcium channel blocker nifedipine (20 μM). Three groups of neurons were studied morphologically: (1) GFP, (2) GFP + GluA1 + vehicle, and (3) GFP + GluA1 + nifedipine. Nifedipine or vehicle was administered daily and cells were prepared for analysis as discussed above (Fig. 2.2). Compared with GFP alone, GluA1 led to an ~40% increase in dendrite branches ($F_{(2,63)} = 8.396$, $p = 0.0006$), ~40% increase in overall arbor size ($F_{(2,63)} = 12.99$, $p < 0.0001$), and ~20 % increase in average dendrite length ($F_{(2,63)} = 9.085$, $p = 0.0004$). Treatment with nifedipine blocked all the prodendrite growth effects of GluA1 overexpression on branching and even suppressed overall tree growth and elaboration of the longest dendrite in comparison with neurons treated with vehicle. It is interesting that recent work has shown that L-type calcium channels play a critical role in homeostatic synaptic plasticity (Goold and Nicoll 2011). Thus, the activity-dependent dendrite growth elicited by GluA1 is NMDA-R independent and requires activation of voltage-gated calcium channels.

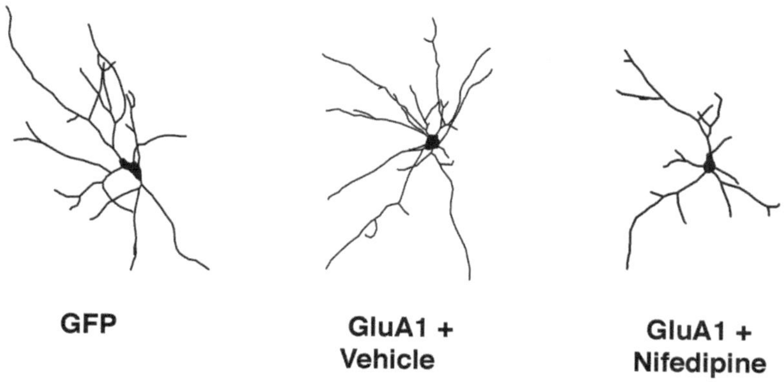

Cell Parameter	GFP Control (n=22)	GluA1 (n=22)	GluA1 and Nifedipine (n=21)	$F_{(2,63)}, p$
Soma area (μm^2)	207.7 ± 50.3	231.9 ± 44.4	213.1 ± 44.5	1.588; 0.2126
1^0 dendrites	6.0 ± 1.0	6.8 ± 1.2	6.3 ± 1.4	2.110; 0.1300
Branches, #	13.6 ± 4.8	18.7 ± 6.2	12.1 ± 5.4†††	8.396; $p = 0.0006$
Σlength (μm)	1125.5 ± 326.1	1542.9 ± 625.2	875.3 ± 242.0†††	12.99; $p < 0.0001$
Average dendrites	192.3 ± 63.7	230.1 ± 89.3	141.6 ± 41.1*†††	9.085; $p = 0.0004$
Longest dendrite (μm)	239.6 ± 82.3	258.8 ± 79.5	177.6 ± 65.5*††	6.531; $p = 0.0027$

*Differs from GFP Control: *p<0.05; ** p<0.01; *** p<0.001;
†Differs from GluA1 Treatment, †p<0.05; †† p<0.01; ††† p<0.001

Fig. 2.2 Overexpression of WT GluA1 stimulates dendrite growth in vitro in a voltage-gated calcium channel-dependent manner. *Top*, representative camera lucida images of neurons expression GFP alone, or GFP + GluA1 (treated or not with nifedipine). The chart *below* provides a quantitative analysis of dendrites as well as a statistical analysis using ANOVA. The number of neurons drawn is noted in parentheses next to the column title. There is statistically significant increase in branching and overall tree size in neurons overexpressing GluA1 and these effects are blocked by voltage-gated calcium channel antagonism

The above-described work argues that expression of GluA1 is sufficient to promote dendrite growth. Simply overexpressing the protein can trigger the elaboration of dendrites (either in vitro or in vivo). To determine if GluA1 is necessary for dendrite growth under endogenous conditions requires that we reduced or eliminated its expression in neurons and determined the effects on morphogenesis. In vitro studies using RNA interference (RNAi) technology reveal that reducing GluA1 in neurons inhibits the normal elaboration of dendrites (Zhang et al. 2008). The effects are dose dependent and cell autonomous. These observations support the view that expression of GluA1 by neurons is necessary for the normal morphogenesis of the dendritic tree.

The In Vivo Role of GluA1 in Motor System Development

Activity-dependent growth of dendrites is a component of experience-dependent motor system development. In light of this, we wondered if the above in vitro observations were also seen in vivo. If so, we would be well positioned to determine the broader impact of GluA1 on other features of experience-dependent motor development such as behavior and circuit connectivity.

GluA1 null mice (GluA1$^{-/-}$) are viable and have defects in cognitive/affective behavioral realms (Bannerman et al. 2004; Zamanillo et al. 1999). We began our studies of these animals by asking if GluA1$^{-/-}$ neonatal motor neurons elaborate a normal dendritic tree. Both at P10 and P23, the dendritic tree of GluA1$^{-/-}$ motor neurons is smaller than wild-type (WT) control mice (Zhang et al. 2008). It is worth noting that in addition to motor neurons, interneurons within the neonatal ventral horn also express GluA1 at very high levels (Jakowec et al. 1995a, b). And so, it is possible that loss of GluA1 from interneurons that innervate motor neurons influences the capacity of motor neurons to elaborate a normal dendritic tree. To examine this issue directly we generated conditional knock-out mice. Mating Hb9-Cre mice with mice bearing a LoxP-flanked allele of GluA1 generates mice with loss of one copy of GluA1. Using an appropriate breeding we generated mice that are homozygous for the LoxP-flanked allele and Cre recombinase and these mice have ablation of GluA1 expression restricted to motor neurons. Analysis of dendrite from these animals ($GluA1^{deltaHb9}$) revealed a reduction of dendrite size and branching, similar but not as severe as what we observed in the GluA1$^{-/-}$ mice (Zhang et al. 2008). There was no effect on dendrite structure in various control mice (i.e., Hb9-Cre alone or GluA1^{LoxP} alone). These in vivo observations support the view that the expression of endogenous GluA1 by motor neurons is required for the normal elaboration of the dendritic tree during development.

Does the loss of GluA1 influence other aspects of the motor system such as circuitry within the segmental spinal cord? To address this question, we used the pseudorabies virus (PRV)-tracing system. A recombinant PRV was generated that expressed GFP in cells. Upon injection into the hamstring leg muscle, PRV-GFP particles are retrogradely transported to motor neurons. PRV-GFP particles are then exported into dendrites where they cross synapses from motor neurons into innervating premotor interneurons. Thus, the distribution of GFP-labeled cells in the spinal cord reflect the pattern of premotor innervation of motor neurons. When we applied this approach to the GluA1$^{-/-}$ and the GluA1$^{+/+}$ mice, we found that much of the GFP labeling was identical between groups. However, fewer labeled interneurons were seen in the ipsilateral Rexed lamina VIII of lumbar segment 4 of the GluA1$^{-/-}$ than the GluA1$^{+/+}$ mice. In addition, there were fewer labeled interneurons in the multiple contralateral Rexed lamina of lumbar segment 2–5 of the GluA1$^{-/-}$ than the GluA1$^{+/+}$ mice (Zhang et al. 2008). These results suggest that segmental spinal cord connectivity is different between the GluA1$^{-/-}$ and the GluA1$^{+/+}$ mice.

Do these alterations in dendrite structure and segmental spinal cord interneuronal connectivity manifest in behavioral differences between the GluA1$^{-/-}$ and the GluA1$^{+/+}$ mice? To examine this, we subjected the two strains of mice to a battery of locomotor tasks including treadmill running, rotarod and fore- and hind-limb grip strength. At P23 and in adulthood, the GluA1$^{-/-}$ mice performed poorer in every single test in comparison with the GluA1$^{+/+}$ mice (Zhang et al. 2008). Similar trends were seen when we studied the $GluA1^{deltaHb9}$ although the degree of locomotor impairment was less than seen in the GluA1$^{-/-}$ mice. These differences in motor function could not be ascribed to a difference in motor neuron number. The weakness phenotype of the GluA1$^{-/-}$ mice was associated with an increase in type I muscle fibers in the gastrocnemius. Thus, elimination of GluA1 from motor neurons (as well as other neurons presumably in the ventral horn) leads to abnormal development of the neuromuscular unit. The dendritic tree of GluA1$^{-/-}$ motor neurons is stunted, the pattern of premotor interneuron connectivity is perturbed, muscle fiber-type specification is distorted and this leads to poorer locomotor performance in comparison with WT animals. These observations point to the critical role that GluA1 plays in the normal activity-dependent development of the motor system.

SAP97 Translates the GluA1-Generated Signal into Dendrite Growth

Our working hypothesis is that synaptic activation of AMPA-Rs assembled with the GluA1 subunit initiate an activity-dependent prodendrite growth signal. Some data suggest that the electrophysiological properties of GluA1-containing AMPA-R regulate how GluA1 influences dendrite morphology, and this is linked to AMPA-R calcium permeability (Jeong et al. 2006). We wondered if, in addition, intracellular proteins that bind GluA1 are also important determinants.

The extreme C-terminal four amino acids of GluA1 act as a ligand for the synapse-associated protein of 97 kDa molecular weight called SAP97 (Cai et al. 2002). SAP97 is a membrane-associated guanylate kinases (MAGUK)-class-scaffolding protein and is the only known binding partner of the extreme C-terminus of GluA1. MAGUK proteins are enriched in the postsynaptic density where they play a variety of roles in synaptic function including chaperoning glutamate receptor subunits into and out of the synapse, receptor clustering and modulation of receptor electrophysiological function (Palmer et al. 2005; Sheng and Sala 2001; Shepherd and Huganir 2007). SAP97 is a modular protein with multiple protein–protein interaction domains. As detailed below, we have explored the dendrite growth-promoting role of SAP97 and its binding to GluA1 in a series of in vitro and vivo experiments.

In co-immunoprecipitation (co-IP) experiments using spinal cord or cerebral cortex tissue, we find that GluA1 and SAP97 are part of a physical complex (Zhou et al. 2008). When the two full-length proteins are expressed in a heterologous system, we can again demonstrate a physical complex in the co-IP assay. Two approaches were taken to establish the portions of each protein required for the physical complex. First, we deleted the C-terminal 7 amino acids of GluA1 (GluA1Δ7). While full-length SAP97 will co-IP full-length GluA1, it will not co-IP GluA1Δ7. Second, the crystal structure of PDZ domains is known and it is possible to introduce mutations such that the PDZ domain becomes incompetent to bind ligands (Morais Cabral et al. 1996). We engineered such mutations into PDZ2 of GluA1 (K323A, K326A) and found that mutant PDZ2 SAP97 did not co-IP full-length GluA1 (Zhou et al. 2008). Thus, GluA1 and SAP97 are part of a physical complex that is likely to be mediated by the binding of the extreme C-terminus of GluA1 to PDZ2 of SAP97.

What biology, if any, is influenced by the GluA1/SAP97 complex? We began by studying the trafficking of GluA1 through the secretory pathway to populate synapses. To address this issue we used a strain of mice in which the WT version of GluA1 has been replaced with a version that is lacking the C-terminal 7 amino acids (Kim et al. 2005). In the homozygous state, this "knock-in" mouse only expresses GluA1Δ7. We find that GluA1Δ7 is synthesized in the endoplasmic reticulum at normal levels, is processed normally in the Golgi apparatus, hetero-oligomerizes normally with other AMPA-R subunits, and inserts into synapses normally (Zhou et al. 2008). Electrophysiological studies of the hippocampus of GluA1Δ7 mice show normal basal synaptic transmission as well as normal LTP/LTP (Kim et al. 2005). Thus, despite the fact that GluA1Δ7 does not physically associate with SAP97, the subunit behaves similar to WT GluA1. In marked contrast, SAP97 does not traffic to synapses in the GluA1Δ7 mice. These observations indicate that GluA1 chaperones SAP97 into synapses.

What is SAP97 doing, in association with GluA1, at synapses? We took two approaches to address this issue. First, we determined the effect on dendrite growth of eliminating SAP97 from neurons. When we knocked down SAP97 with a small hairpin RNA (shRNA), the neuronal dendritic tree is smaller and less branched than WT neurons (Zhou et al. 2008). This implies that endogenous SAP97 is required for normal elaboration of the dendritic tree. We also found that knockdown of SAP97 blocked the dendrite growth-promoting action of GluA1 overexpression. To validate these in vitro observations, we wanted to study mice in which SAP97 is ablated. Unfortunately, SAP97 null mice die at birth owing to cranio-facial abnormalities. To overcome this problem we generated mice in which SAP97 is eliminated specifically in motor neurons. This was achieved using the Hb9-Cre mice mated to mice bearing a floxed allele of SAP97. We found that the dendrites of motor neuron from the $SAP97^{deltaHb9}$ mice are smaller and less branched than WT mice (Zhou et al. 2008). Thus, both in vitro and in vivo studies demonstrate that SAP97 plays a key role in the normal development of the neuronal dendritic tree. In addition, all of the dendrite growth-promoting actions of GluA1 are lost in the absence of SAP97. This suggests that SAP97 acts to translate activity from GluA1-containing AMPA-Rs into growth.

Our second approach to understanding what GluA1 and SAP97 are doing at synapses focused on the nature of their physical relationship. Must SAP97 be physically tethered to GluA1 to promote dendrite growth? Or, is colocalization of both proteins to the plasma membrane sufficient for GluA1 to promote dendrite growth? We undertook a series of experiments to explore this issue. First, we found that GluA1, but not GluA1Δ7, overexpressed in neurons in vitro is dendrite growth promoting (Zhou et al. 2008). In addition, we found that coexpression of GluA1 with SAP97 has a synergistic dendrite growth-promoting action, while coexpression of GluA1Δ7 with SAP97 leads to modest dendrite growth (equivalent to the dendrite growth-promoting action of SAP97 itself). So, even though GluA1Δ7 traffics to the cell surface and hetero-oligomerizes normally with other AMPA-R subunits, the lack of physical association with SAP97 blocks the dendrite growth-promoting action of this subunit.

In the next set of experiments we added a palmitoylation sequence to SAP97 (palSAP97) and we show that this leads to membrane targeting of the protein. Both SAP97 and palSAP97 have equivalent dendrite growth-promoting actions when overexpressed in neurons. Armed with this tool we took two approaches to look at the necessity of a physical interaction between GluA1 and SAP97 for the promotion of dendrite growth. First, we asked if coexpression of palSAP97 with GluA1Δ7 rescued the dendrite growth-promoting activity of this version of GluA1. Remarkably, the combination of palSAP97 with GluA1Δ7 promoted dendrite growth to the same degree that the combination of SAP97 + GluA1 did (Zhou et al. 2008).

In our final set of in vitro studies, we palmitoylated the version of SAP97 that contained mutations in PDZ2 that disrupted its physical association with GluA1 (mutPDZ2-palSAP97). In coexpression studies in heterologous cells, we found that GluA1 will co-IP palSAP97, but will not co-IP mutPDZ2-palSAP97 (Zhou et al. 2008). So, even though mutPDZ2-palSAP97 targets to the plasma membrane, this is not sufficient to lead to a physical association with GluA1. We then asked about the dendrite growth-promoting action of mutPDZ2-palSAP97 and we found that when coexpressed with GluA1, both palSAP97 and mutPDZ2-palSAP97 were equally effective in promoting dendrite growth (Zhou et al. 2008). Thus, using two different strategies to disrupt the physical association of SAP97 with GluA1, we come to the same conclusion: coexpression of SAP97 with GluA1 synergistically promoted dendrite growth as long as both proteins are targeted to the plasma membrane. While the native proteins associate as part of a physical complex, experimental manipulations that delink the two proteins demonstrate that colocalization, not physical interaction, are required for dendrite growth.

Potential Implications

Why should we care about this pathway of activity-dependent neuronal plasticity? One reason is that knowledge of this pathway may lead to ways of promoting plasticity in adults. One potential beneficiary might be individuals with spinal cord injury (Kalb 2003). After a thoracic spinal cord lesion, the circuitry in the lumbar spinal cord can be engaged by repetitive activation of selected neuronal pathways (e.g., standing training, ambulation training) which results in remarkable improvement in motor behavior. This is seen both in experimental animals and humans (Barbeau and Rossignol 1987; Dietz et al. 1995; Edgerton et al. 1997, 2004; Fung et al. 1990; Lovely et al. 1990; Rossignol 2000; Wernig et al. 1995, 1998; Wirz et al. 2001). The mechanism for this effect is an use-dependent modification of spinal cord circuitry (Gazula et al. 2004) and so we think that enhancement of activity-dependent plasticity within the spinal cord will have a salubrious effect on functional recovery.

Another reason to study this form of activity-dependent neuronal plasticity relates to developmental disorders of brain. Abnormalities in dendrite structure (i.e., size, branching, and spines) are commonly seen in childhood diseases such as mental retardation, autism, and autism-spectrum disorders (Dierssen and Ramakers 2006;

Kaufmann and Moser 2000). Several lines of evidence indicate that in some forms of these childhood diseases, the primary defect is in activity-dependent development. Many genes linked to familial forms of impaired cognitive and emotional development are involved in activity-dependent synapse formation or stabilization (i.e., actin-related proteins such as cofilin, LIMK, and debrin; Rho-GTPase regulators such as oligophrenin-1 and Kalirin-7; and trophic factors such as BDNF, NRGN-ErbB4; Lin and Koleske 2010). In a study using homozygosity mapping to discover recessive disease genes in autistic patients (Morrow et al. 2008), significant genetic heterogeneity was found. One of the more remarkable findings of this study was that many autism-associated genes are regulated by neuronal activity. For example, the expression of the candidate gene DIA1 is regulated by activity and this transcription factor that controls the expression of other activity-regulated transcripts such as MEF2, NPAS4, CREB, EGR, SRF, and others. If we start with the proposition that perturbation of experience-dependent cortical development underlies some of the defects in autism, then it is critical to understand the varieties of normal activity-dependent development. In this regard, it is perhaps noteworthy that genetic studies link SAP97 to schizophrenia (Sato et al. 2008; Toyooka et al. 2002) and autism (Willatt et al. 2005). It is possible that exploration of motor system development will provide a window onto previously unknown aspects of brain operation.

Acknowledgments This work was supported in the past by the US Public Health Service (NS29837). We thank R. Sprengel, P Seeburg, and R. Huganir for several of the murine strains used in these studies.

References

Aizawa H, Hu SC, Bobb K, Balakrishnan K, Ince G, Gurevich I et al (2004) Dendrite development regulated by CREST, a calcium-regulated transcriptional activator. Science 303(5655):197–202

Altman J, Sudarshan K (1975) Postnatal development of locomotion in the laboratory rat. Animal Behavior 23:896–920

Bannerman DM, Deacon RM, Brady S, Bruce A, Sprengel R, Seeburg PH et al (2004) A comparison of GluR-A-deficient and wild-type mice on a test battery assessing sensorimotor, affective, and cognitive behaviors. Behav Neurosci 118(3):643–647

Barbeau H, Rossignol S (1987) Recovery of locomotion after chronic spinalization in the adult cat. Brain Res 412:84–95

Bi G, Poo M (2001) Synaptic modification by correlated activity: Hebb's postulate revisited. Annu Rev Neurosci 24:139–166

Cai C, Coleman SK, Niemi K, Keinanen K (2002) Selective binding of synapse-associated protein 97 to GluR-A alpha-amino-5-hydroxy-3-methyl-4-isoxazole propionate receptor subunit is determined by a novel sequence motif. J Biol Chem 277(35):31484–31490

Carriedo SG, Yin HZ, Weiss JH (1996) Motor neurons are selectively vulnerable to AMPA/KA receptor-mediated injury in vitro. J Neurosci 16:4069–4079

Chen SX, Tari PK, She K, Haas K (2010) Neurexin-neuroligin cell adhesion complexes contribute to synaptotropic dendritogenesis via growth stabilization mechanisms in vivo. Neuron 67(6): 967–983

Chklovskii DB (2004) Synaptic connectivity and neuronal morphology: two sides of the same coin. Neuron 43(5):609–617

Cline HT (2001) Dendritic arbor development and synaptogenesis. Curr Opin Neurobiol 11(1): 118–126
Cline H, Haas K (2008) The regulation of dendritic arbor development and plasticity by glutamatergic synaptic input: a review of the synaptotrophic hypothesis. J Physiol 586(6):1509–1517
Cline HT, Debski EA, Constantine-Paton M (1987) N-methyl-D-aspartate receptor antagonist desegregates eye-specific columns. PNAS 84:4342–4345
Constantine-Paton M, Cline HT, Debski E (1990) Patterned activity, synaptic convergence, and the NMDA receptor in developing visual pathways. Annu Rev Neurosci 13:129–154
Cramer KS, Angelucci A, Hahm JO, Bogdanov MB, Sur M (1996) A role for nitric oxide in the development of the ferret retinogeniculate projection. J Neurosci 16:7995–8004
Curfs MHJM, Gribnau AAM, Dideren PJWC (1993) Postnatal maturation of the dendritic fields of motoneuron pools supplying flexor and extensor muscles of the distal forelimb in the rat. Development 117:535–541
Curfs MHJM, Gribnau AAM, Dederen PJWC (1994) Selective elimination of transient corticospinal projections in the rat cervical spinal cord gray matter. Dev Brain Res 78:182–190
Dierssen M, Ramakers GJ (2006) Dendritic pathology in mental retardation: from molecular genetics to neurobiology. Genes Brain Behav 5(Suppl 2):48–60
Dietz V, Colombo G, Jensen L, Baumgartner L (1995) Locomotor capacity of spinal cord in paraplegic patients. Ann Neurol 37(5):574–582
Donatelle JM (1977) Growth of the Corticospinal tract and the development of placing reactions in the postnatal rat. J Comp Neurol 175:207–232
Edgerton VR, de Leon RD, Tillakaratne N, Recktenwald MR, Hodgson JA, Roy RR (1997) Use-dependent plasticity in spinal stepping and standing. Adv Neurol 72:233–247
Edgerton VR, Tillakaratne NJ, Bigbee AJ, de Leon RD, Roy RR (2004) Plasticity of the spinal neural circuitry after injury. Annu Rev Neurosci 27:145–167
Engert F, Tao HW, Zhang LI, Poo MM (2002) Moving visual stimuli rapidly induce direction sensitivity of developing tectal neurons. Nature 419(6906):470–475
Fung J, Stewart JE, Barbeau H (1990) The combined effects of clonidine and cyproheptadine with interactive training on the modulation of locomotion in spinal cord injured subjects. J Neurol Sci 100(1–2):85–93
Gao FB, Bogert BA (2003) Genetic control of dendritic morphogenesis in Drosophila. Trends Neurosci 26(5):262–268
Gaudilliere B, Konishi Y, de la Iglesia N, Yao G, Bonni A (2004) A CaMKII-NeuroD signaling pathway specifies dendritic morphogenesis. Neuron 41(2):229–241
Gazula VR, Roberts M, Luzzio C, Jawad AF, Kalb RG (2004) Effects of limb exercise after spinal cord injury on motor neuron dendrite structure. J Comp Neurol 476(2):130–145
Goodman CS, Shatz CJ (1993) Developmental mechanisms that generate precise patterns of neuronal connectivity. Cell 10:77–98
Goold CP, Nicoll RA (2011) Single-cell optogenetic excitation drives homeostatic synaptic depression. Neuron 68(3):512–528
Ha S, Redmond L (2008) ERK mediates activity dependent neuronal complexity via sustained activity and CREB-mediated signaling. Dev Neurobiol 68(14):1565–1579
Hume RI, Purves D (1981) Geometry of neonatal neurons and the regulation of synapse elimination. Nature 293:469–471
Inglis FM, Furia F, Zuckerman KE, Strittmatter SM, Kalb RG (1998) The role of nitric oxide and NMDA receptors in the development of motor neuron dendrites. J Neurosci 18:10493–10501
Inglis FM, Zuckerman KE, Kalb RG (2000) Experience-dependent development of spinal motor neurons. Neuron 26:299–305
Inglis FM, Crockett R, Korada S, Abraham WC, Hollmann M, Kalb RG (2002) The AMPA receptor GluR1 regulates dendritic architecture of motor neurons. J Neurosci 22:8042–8051
Jakowec MW, Fox AJ, Martin LJ, Kalb RG (1995a) Quantitative and Qualitative Changes in AMPA Receptor Expression During Spinal Cord Development. Neuroscience 67:893–907
Jakowec MW, Yen L, Kalb RG (1995b) In situ hybridization analysis of AMPA receptor subunit gene expression in the developing rat spinal cord. Neuroscience 67:909–920

Jan YN, Jan LY (2003) The control of dendrite development. Neuron 40(2):229–242
Jaworski J, Spangler S, Seeburg DP, Hoogenraad CC, Sheng M (2005) Control of dendritic arborization by the phosphoinositide-3′-kinase-Akt-mammalian target of rapamycin pathway. J Neurosci 25(49):11300–11312
Jeong GB, Werner M, Gazula VR, Itoh T, Roberts M, David S et al (2006) Bi-directional control of motor neuron dendrite remodeling by the calcium permeability of AMPA receptors. Mol Cell Neurosci 32(3):299–314
Kalb RG (1994) Regulation of motor neuron dendrite growth by NMDA receptor activation. Development 120:3063–3071
Kalb RG (2003) Getting the spinal cord to think for itself. Arch Neurol 60(6):805–808
Katz LC, Constantine-Paton M (1988) Relationships between segregated afferents and postsynaptic neurons in the optic tectum of three-eyed frogs. J Neurosci 8:3160–3180
Kaufmann WE, Moser HW (2000) Dendritic anomalies in disorders associated with mental retardation. Cereb Cortex 10(10):981–991
Kim CH, Takamiya K, Petralia RS, Sattler R, Yu S, Zhou W et al (2005) Persistent hippocampal CA1 LTP in mice lacking the C-terminal PDZ ligand of GluR1. Nat Neurosci 8(8):985–987
Kleinschmidt A, Bear MF, Singer W (1987) Blockade of "NMDA" receptors disrupts experience-dependent plasticity of kitten striate cortex. Science 238:355–358
Kumar V, Zhang MX, Swank MW, Kunz J, Wu GY (2005) Regulation of dendritic morphogenesis by Ras-PI3K-Akt-mTOR and Ras-MAPK signaling pathways. J Neurosci 25(49):11288–11299
Li Z, Van Aelst L, Cline HT (2000) Rho GTPases regulate distinct aspects of dendritic arbor growth in Xenopus central neurons in vivo. Nat Neurosci 3(3):217–225
Li Z, Aizenman CD, Cline HT (2002) Regulation of rho GTPases by crosstalk and neuronal activity in vivo. Neuron 33(5):741–750
Li S, Zhang C, Takemori H, Zhou Y, Xiong ZQ (2009) TORC1 regulates activity-dependent CREB-target gene transcription and dendritic growth of developing cortical neurons. J Neurosci 29(8):2334–2343
Lin YC, Koleske AJ (2010) Mechanisms of synapse and dendrite maintenance and their disruption in psychiatric and neurodegenerative disorders. Annu Rev Neurosci 33:349–378
Lindsay AD, Greer JJ, Feldman JL (1991) Phrenic Motoneuron Morphology in the Neonatal Rat. J Comp Neurol 308:169–179
Lovely RG, Gregor RJ, Roy RR, Edgerton VR (1990) Weight-bearing hindlimb stepping in treadmill-exercised adult spinal cats. Brain Res 514:206–218
Lu B (2003) Pro-region of neurotrophins: role in synaptic modulation. Neuron 39(5):735–738
McAllister AK, Katz LC, Lo DC (1996) Neurotrophin regulation of cortical dendritic growth requires activity. Neuron 17(6):1057–1064
McAllister AK, Katz LC, Lo DC (1997) Opposing roles for endogenous BDNF and NT-3 in regulating cortical dendrite growth. Neuron 18:767–778
Morais Cabral JH, Petosa C, Sutcliffe MJ, Raza S, Byron O, Poy F et al (1996) Crystal structure of a PDZ domain. Nature 382(6592):649–652
Morrow EM, Yoo SY, Flavell SW, Kim TK, Lin Y, Hill RS et al (2008) Identifying autism loci and genes by tracing recent shared ancestry. Science 321(5886):218–223
Núez-Abades PA, He F, Barrionuevo G, Cameron WE (1994) Morphology of developing rat genioglossal motoneurons studied in vitro: changes in length, branching pattern, and spatial distribution of dendrites. J Comp Neurol 339:401–420
Palmer CL, Cotton L, Henley JM (2005) The molecular pharmacology and cell biology of alpha-amino-3-hydroxy-5-methyl-4-isoxazolepropionic acid receptors. Pharmacol Rev 57(2): 253–277
Pellis V, Pellis S, Teitelbaum P (1991) A descriptive analysis of the postnatal development of contact-righting in rats (Rattus norvegicus). Devel Psychobiol 24(4):237–263
Peng YR, He S, Marie H, Zeng SY, Ma J, Tan ZJ et al (2009) Coordinated changes in dendritic arborization and synaptic strength during neural circuit development. Neuron 61(1):71–84
Redmond L, Oh SR, Hicks C, Weinmaster G, Ghosh A (2000) Nuclear Notch1 signaling and the regulation of dendritic development. Nat Neurosci 3(1):30–40

Redmond L, Kashani AH, Ghosh A (2002) Calcium regulation of dendritic growth via CaM kinase IV and CREB-mediated transcription. Neuron 34(6):999–1010
Rossignol S (2000) Locomotion and its recovery after spinal injury. Curr Opin Neurobiol 10: 708–716
Ruthazer ES, Akerman CJ, Cline HT (2003) Control of axon branch dynamics by correlated activity in vivo. Science 301(5629):66–70
Sato J, Shimazu D, Yamamoto N, Nishikawa T (2008) An association analysis of synapse-associated protein 97 (SAP97) gene in schizophrenia. J Neural Transm 115(9):1355–1365
Seebach BS, Ziskind-Conhaim L (1994) Formation of transient inappropriate sensorimotor synapses in developing rat spinal cord. J Neurosci 14:4520–4528
Shalizi A, Gaudilliere B, Yuan Z, Stegmuller J, Shirogane T, Ge Q et al (2006) A calcium-regulated MEF2 sumoylation switch controls postsynaptic differentiation. Science 311(5763):1012–1017
Shatz CJ (1990) Impulse activity and the patterning of connections during CNS development. Neuron 5:745–756
Sheng M, Sala C (2001) PDZ domains and the organization of supramolecular complexes. Annu Rev Neurosci 24:1–29
Shepherd JD, Huganir RL (2007) The cell biology of synaptic plasticity: AMPA receptor trafficking. Annu Rev Cell Dev Biol 23:613–643
Sin WC, Haas K, Ruthazer ES, Cline HT (2002) Dendrite growth increased by visual activity requires NMDA receptor and Rho GTPases. Nature 419(6906):475–480
Snider WD, Zhang L, Yusoof S, Gorukanti N, Tsering C (1992) Interactions between dorsal root axons and their target motor neurons in developing mammalian spinal cord. J Neurosci 12(9):3494–3508
Stegenga SL, Kalb RG (2001) Developmental regulation of N-methyl-D-aspartate: and kainate-type glutamate receptor expression in the rat spinal cord. Neuroscience 105:499–507
Stuart G, Spruston N, Hausser M (1999) Dendrites. Oxford: Oxford University Press
Thyagarajan A, Ting AY (2010) Imaging activity-dependent regulation of neurexin-neuroligin interactions using trans-synaptic enzymatic biotinylation. Cell 143(3):456–469
Toyooka K, Iritani S, Makifuchi T, Shirakawa O, Kitamura N, Maeda K et al (2002) Selective reduction of a PDZ protein, SAP-97, in the prefrontal cortex of patients with chronic schizophrenia. J Neurochem 83(4):797–806
Vandenberghe W, Robberecht W, Brorson JR (2000) AMPA receptor calcium permeability, GluR2 expression, and selective motoneuron vulnerability. J Neurosci 20(1):123–132
Vaughn JE (1989) Fine structure of synaptogenesis in the vertebrate central nervous system. Synapse 3:255–285
Vaughn JE, Barber RP, Sims TJ (1988) Dendritic development and preferential growth into synaptogenic fields: a quantitative study of Golgi-impregnated spinal motor neurons. Synapse 2(1):69–78
Walton KD, Lieberman D, Llinas A, Begin M, Llinas RR (1992) Identification of a critical period for motor development in neonatal rats. Neuroscience 51:763–767
Walton KD, Harding S, Anschel D, Harris YT, Llinas R (2005) The effects of microgravity on the development of surface righting in rats. J Physiol 565:593–608
Wayman GA, Impey S, Marks D, Saneyoshi T, Grant WF, Derkach V et al (2006) Activity-dependent dendritic arborization mediated by CaM-kinase I activation and enhanced CREB-dependent transcription of Wnt-2. Neuron 50(6):897–909
Wernig A, Muller S, Nanassy A, Cagol E (1995) Laufband therapy based on "rules of spinal locomotion" is effective in spinal cord injured persons. Eur J Neurosci 7:823–829
Wernig A, Nanassy A, Muller S (1998) Maintainance of locomotor ability following Laufband (treadmill) therapy in para- and tetraplegia persons: follow-up studies. Spinal Cord 36:744–749
Willatt L, Cox J, Barber J, Cabanas ED, Collins A, Donnai D et al (2005) 3q29 microdeletion syndrome: clinical and molecular characterization of a new syndrome. Am J Hum Genet 77(1):154–160
Wirz M, Colombo G, Dietz V (2001) Long term effects of locomotor training in spinal humans. J Neurol Neurosurg Psychiatry 71(1):93–96

Wu GY, Cline HT (1998) Stabilization of dendritic arbor structure in vivo by CaMKII. Science 279:222–225

Wu HH, Williams CV, McLoon SC (1994) Involvement of nitric oxide in the elimination of a transient retinotectal projection in development. Science 265:1593–1596

Wu GY, Deisseroth K, Tsien RW (2001) Spaced stimuli stabilize MAPK pathway activation and its effects on dendritic morphology. Nat Neurosci 4(2):151–158

Yu X, Malenka RC (2003) Beta-catenin is critical for dendritic morphogenesis. Nat Neurosci 6(11):1169–1177

Zamanillo D, Sprengel R, Hvalby O, Hensen V, Burnashev N, Rozov A et al (1999) Importance of AMPA receptors for hippocampal synaptic plasticity but not spatial learning. Science 284:1805–1811

Zhang LI, Tao HW, Poo M (2000) Visual input induces long-term potentiation of developing retinotectal synapses. Nat Neurosci 3(7):708–715

Zhang L, Schessl J, Werner M, Bonnemann C, Xiong G, Mojsilovic-Petrovic J et al (2008) Role of GluR1 in activity-dependent motor system development. J Neurosci 28(40):9953–9968

Zhou W, Zhang L, Guoxiang X, Mojsilovic-Petrovic J, Takamaya K, Sattler R et al (2008) GluR1 controls dendrite growth through its binding partner, SAP97. J Neurosci 28(41):10220–10233

Zou DJ, Cline HT (1999) Postsynaptic calcium/calmodulin-dependent protein kinase II required to limit elaboration of presynaptic and postsynaptic neuronal arbors. J Neurosci 19:8909–8919

Chapter 3
Neurocognitive Mechanisms of Error-Based Motor Learning

Rachael D. Seidler, Youngbin Kwak, Brett W. Fling and Jessica A. Bernard

Introduction

During the initial stages of skill learning, motor performance is cognitively demanding and uncoordinated. Understanding how an individual progresses to the exquisite and automatic processes of proficient performance has not been a trivial process. Here, we review the role that error detection and correction plays in skill learning. We make a distinction between *within trial* error corrections, or performance adjustments made during the course of a movement that is not achieving its intended goal, and *across trial* error corrections, or cumulative adjustments that reflect ongoing learning. We review recent debates regarding whether skill learning is a purely implicit or subconscious process, or if it can benefit from explicit instruction.

Learning from errors is one of the basic principles of motor skill acquisition. Current ideas about error-based learning stem from forward model control theories (Diedrichsen et al. 2010; Miall and Wolpert 1996). When movement errors are

R. D. Seidler (✉)
Department of Psychology and School of Kinesiology, University of Michigan, 401 Washtenaw Avenue, Ann Arbor, MI 48109-2214, USA
e-mail: rseidler@umich.edu

Y. Kwak
Neuroscience Program, University of Michigan, 401 Washtenaw Avenue, Ann Arbor, MI 48109-2214, USA
e-mail: youngbin.kwak@duke.edu

Center for Cognitive Neuroscience, Duke University, Durham, NC 27708, USA

B. W. Fling
School of Kinesiology, University of Michigan, 401 Washtenaw Avenue, Ann Arbor, MI 48109-2214, USA
e-mail: bfling@hs.uci.edu

J. A. Bernard
Department of Psychology, University of Michigan, 401 Washtenaw Avenue, Ann Arbor, MI 48109-2214, USA
e-mail: jessbern@umich.edu

M. J. Richardson et al. (eds.), *Progress in Motor Control,*
Advances in Experimental Medicine and Biology 782,
DOI 10.1007/978-1-4614-5465-6_3,

detected by sensory systems, the information is used to update the motor commands for subsequent actions. However, relying solely on sensory feedback does not allow efficient motor adjustments because of the time delay between the initial motor command and the arrival of sensory feedback. Movement induces continuous changes to state variables such as limb position and velocity. In order to allow accurate movement adjustments, the motor system relies on a forward model that makes predictions of the sensory outcomes (i.e., changes in position and velocity) associated with a given motor command (Bastian 2006; Flanagan et al. 2003). Differences between the predicted and actual sensory outcome serve as the feedback error signal that updates forthcoming motor commands.

When learning a new motor skill such as swinging a golf club, new skills do not have enough of a motor history for an accurate forward model, resulting in large prediction errors. In this case, the process of learning involves updating motor commands through multiple exposures to motor errors and gradually reducing them by refining the forward model (Donchin et al. 2003; Shadmehr et al. 2010).

The mechanisms of error-based learning are often studied using visuomotor adaptation and force field adaptation tasks. Visuomotor adaptation involves distortion of the visual consequences of movement, whereas force field adaptation affects the proprioceptive consequences of motor commands by altering the dynamics of movement (Shadmehr and Mussa-Ivaldi 1994). Error processing under these two paradigms shows extensive neural overlap in the cerebellum, suggesting a common mechanism for error processing and learning (Diedrichsen et al. 2005).

While extensive evidence supports a role for the cerebellum in error-based sensorimotor learning (Criscimanga-Hemminger et al. 2010; Diedrichsen et al. 2005; Ito 2002; Miall and Wolpert 1996; Miall et al. 1993, 2007; Ramnani 2006; Tseng et al. 2007), many neuroimaging studies also provide evidence that brain regions other than the cerebellum may play a role, including the parietal cortex, striatum and anterior cingulate cortex (Clower et al. 1996; Danckert et al. 2008; den Ouden et al. 2010). In this chapter, we review the literature supporting the involvement of three distinct brain structures in error processing: (1) the anterior cingulate cortex (ACC), (2) the basal ganglia, and (3) the cerebellum. We also speculate about the specific roles that each structure may play during motor learning, depending on the learning task and context. In an effort to add some structure to the burgeoning literature on this topic, we have organized our review into discussions of both feedforward and feedback learning processes (cf. Wolpert et al. 1998).

Anterior Cingulate/Medial Frontal Cortex Contributions to Error Processing

One of the brain systems that plays a critical role in error processing is the medial prefrontal/ACC region (see Fig. 3.1; Seidler et al. in press). This prefrontal performance monitoring system has been studied extensively by recording the error-related negativity (ERN), an event-related potential (ERP) component that is time locked

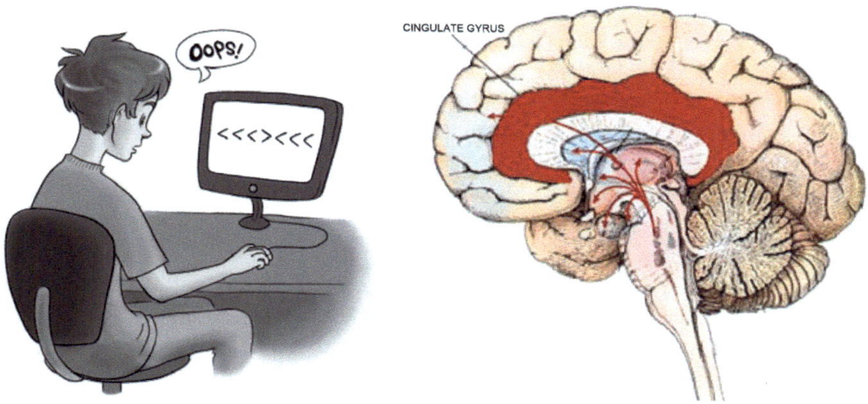

Fig. 3.1 The *left* panel illustrates a participant performing the Flanker task, classically employed to study the event related negativity (ERN) associated with error commission. The task goal is to respond to the direction of the center *arrow* while ignoring the conflicting information represented in the surrounding areas (art by Lauren Wu). The *right* panel indicates the cingulate gyrus; it is thought that the ERN originates in the anterior cingulate gyrus

to an erroneous response (Falkenstein et al. 1995; Gehring et al. 1993, 1995). The ERN is thought to be generated in the anterior cingulate cortex (ACC), which is known to serve cognitive control functions that enable the brain to adapt behavior to changing task demands and environmental circumstances (Botvinick et al. 2001; Ridderinkhof et al. 2004). The cognitive control and performance monitoring functions of this brain region have been predominantly studied using cognitive tasks such as the Stroop word-color interference task and Eriksen's flanker task (Nachev et al. 2008), depicted in Fig. 3.1. These tasks require individuals to minimize interference from irrelevant, conflicting cues, and to monitor incorrect trials to adjust and improve performance. In the following section, we review recent literature demonstrating how the ACC/medial prefrontal system contributes to feedback and feedforward motor learning.

Feedback Error Processing in the ACC

The ACC error monitoring system has traditionally been viewed as contributing to the feedback processing of errors. The ERN is a response-locked ERP component that appears *after* commission of an error, which is monitored by the medial prefrontal/ACC system and contributes to performance improvement (Gehring et al. 2011). This is also represented by the reinforcement learning theory that explains the origin of the ERN and the error monitoring role of the medial prefrontal/ACC region (Holroyd and Coles 2002). This theory suggests that the medial prefrontal/ACC receives reward prediction error signals from midbrain dopaminergic cells, which also send the same input to the basal ganglia (described in more detail in the subsequent section). When the error signal is delivered to the ACC, this system contributes to performance improvement by exerting greater cognitive control (Gehring et al. 2011). Another theory that explains the mechanism of feedback processing of ERN is the error detection/comparator theory (Gehring et al. 2011). According to this theory, the ERN is a signal representing the mismatch between the actual output of the motor system and the best estimate of the correct response (Falkenstein et al. 1991; Gehring et al. 1993). This mismatch signal is conveyed to the control center for future motor command adjustment as part of the feedback process.

While traditionally ERN feedback processing has been studied with cognitive tasks, recent work also demonstrates a contribution of the medial prefrontal/ACC error processing system to motor control and motor learning (Anguera et al. 2010; Anguera et al. 2009; Danckert et al. 2008; Ferdinand et al. 2008).

We recently tested whether the ERN was sensitive to the magnitude of error experienced during a visuomotor adaptation task and found a larger ERN on trials in which larger motor errors were made (Anguera et al. 2009). ERN magnitude also decreased from the early to the late stages of learning. These results are in agreement with current theories of the ERN and skill acquisition. For example, as the error detection theory proposes (Falkenstein et al. 1991; Gehring et al. 1993), a greater ERN associated with larger errors indicates that the brain was monitoring the disparity between the predicted and actual movement outcomes (Anguera et al. 2009).

There is also evidence supporting the notion that error processing in the ACC contributes to motor sequence learning (Berns et al. 1997). The N200 ERP component, which is also localized in ACC, is known to be sensitive to response conflict and cognitive control (Folstein and Van Petten 2008). The N200 component has also been widely studied in the domain of feedback or error monitoring together with the ERN (Folestein and Van Petten 2008; Gehring et al. 2011). Several studies have shown that the N200 is enhanced for a stimulus that violates a learned motor sequence (Eimer et al. 1996). Similarly, when ERN magnitudes were compared between explicit and implicit sequence learners, a larger ERN was found for the explicit learners demonstrating greater involvement of the error monitoring system when individuals are actively searching for the regularity of a sequence (Russeler et al. 2003). A more recent study demonstrated a parametric increase in the magnitude of the ERN during sequence learning as the awareness of the sequential nature and the predictability of the forthcoming sequential element increased (Ferdinand et al. 2008).

Interestingly, a number of EEG source localization studies have suggested that motor regions such as the cingulate motor area and the presupplementary motor area (pre-SMA) are the generator sites of the ERN as opposed to the ACC (Dhar and Pourtois 2011; Hochman et al. 2009; Badgaiyan and Posner 1998; Dehaene et al. 1994; Miltner et al. 1997). These motor regions not only take part in higher order motor executive control and self-initiated movements, but also contribute to sequential movements and adaptive motor learning (Chao et al. 2009; Duann et al. 2009; Stuphorn et al. 2010; Chen et al. 2010; Nachev et al. 2005; Cunnington et al. 2002; Shima et al. 1996; Chen and Wise 1996; Hikosaka et al. 1996). Clearly further investigation is required to parse out the functions of these individual medial brain regions to error-based learning.

Feedforward Error Processing in the ACC

While the role of the ACC in error processing has historically been more focused on feedback processes, recent work suggests that medial prefrontal regions including the ACC and medial motor areas also serve feedforward functions during motor tasks. One example is shown during motor response inhibition, as measured by the stop signal task. During this task, participants are asked to cancel their prepotent motor response when they see a stop signal that occurs occasionally, is unpredictable, and occurs at various latencies after the appearance of the target stimulus (Logan 1994). Stop signal response time is an estimation of the time an individual needs to inhibit a prepared motor response (Logan 1994). Studies have shown that the pre-SMA together with the inferior frontal gyrus is involved in stop signal task performance (Chao et al. 2009; Duann et al. 2009). In particular, successful motor response inhibition is represented as shorter stop signal response times, and activates the pre-SMA (Chao et al. 2009). The role that this motor region plays in higher order motor control is interpreted as a proactive control system that allows intertrial adjustments to the level of motor readiness based on prior performance and anticipated task requirements (Stuphorn et al. 2010; Chen et al. 2010). This suggests that the medial motor regions also serve a feedforward error processing role during complex motor behaviors. Whether the error processing mechanisms of the cingulate motor area and the pre-SMA are distinct from those of the medial prefrontal/ACC regions during cognitive control tasks is debatable. Considering the relative anatomical closeness of these regions and the evidence suggesting that the cingulate motor area and the pre-SMA may also be generator sites of the ERN signal (Nachev et al. 2008; Dhar and Pourtois 2011; Hochman et al. 2009; Badgaiyan and Posner 1998; Dehaene et al. 1994; Miltner et al. 1997), one might argue that these regions serve similar functions. However, it has been proposed that the cingulate motor area and the pre-SMA correct for movement errors in a proactive manner (Isoda and Hikosaka 2007).

In a series of studies, Krigolson et al. have demonstrated a contribution of the ERN to feedforward error processing in a joystick movement motor task (Krigolson and Holroyd 2006, 2007a, 2007b; Krigolson et al. 2008). They found that the ERN began just prior to target tracking errors, indicating that the medial frontal system began to detect the error even before it was fully committed (Krigolson and Holroyd 2006). The authors suggested that this might entail the medial frontal system predicting tracking errors by adopting a predictive mode of control (Desmurget et al. 2000). That is, these results indicate a feedforward role of the medial frontal ERN to motor error processing that is distinct from error feedback processes (Krigolson and Holroyd 2007a; Krigolson et al. 2008).

In conclusion, studies have provided support that the medial prefrontal/ACC system, together with the cingulate motor areas and the pre-SMA play a role in motor error processing and skill acquisition. However, the actual mechanisms whereby these two systems contribute to performance improvements across trials are not well understood. It is also unclear whether this system works independently of or in collaboration with the cerebellum and basal ganglia error processing systems, although it is likely that this system works in concert with the basal ganglia networks as both regions receive similar midbrain dopaminergic inputs as described above.

Fig. 3.2 The cartoons depict rewarding stimuli (*left* and *middle* panels, art by Lauren Wu), while the *right* panel highlights the basal ganglia, known to be involved in processing error and reward information in reinforcement learning paradigms

Basal Ganglia Contributions to Error Processing

Errors play an important role in goal-directed behavior. When the consequences of our behavior are better than expected, associations and response patterns are strengthened (Hebb 1949). Conversely, when outcomes are worse than expected (e.g., when performance errors occur), adjustment often occurs resulting in improved performance (Rabbitt 1966).

Empirical and computational work implicates the midbrain dopamine (DA) system and its most prominent target, the basal ganglia (see Fig. 3.2), as key sub-cortical structures involved in error processing (Schultz 1998, 2001). Dopaminergic projections are most highly concentrated in the dorsal striatum and the ventral bank of the ACC (Goldman-Rakic et al. 1989; Smith and Bolam 1990). Dopamine neurons also receive inputs, primarily from the ventral striatum and caudal orbitofrontal cortex (Eblen and Graybiel 1995). These projections exert control over dopamine output, which can modulate stimulus–response associations (Horvitz 2002). For example, performance errors initiate neural training signals that alter our response tendencies (Holroyd and Coles 2002; Ljungberg et al. 1991). Errors briefly deactivate midbrain dopamine neurons (for approximately 100 ms following error commission), which carry predictive error signals to various parts of the brain for reinforcement learning (Guigon et al. 1995). Changes in basal ganglia activity modulate the thalamus, which then relays information to the cortex.

There are several functional MRI (fMRI) studies which suggest a role for the human basal ganglia (BG) in regulating many kinematic properties. For example, activity in the putamen and the internal globus pallidus (GPi) is associated with movement velocity and amplitude (Turner et al. 1998, 2003; Spraker et al. 2007), putamen activity also scales with force duration (Vaillancourt et al. 2004; Prodoehl et al. 2008), and GPi and subthalamic nucleus activation is related to force rate of change and amplitude (Spraker et al. 2007; Prodoehl et al. 2008). There is also a large body of fMRI studies showing error-related increases in blood-oxygen level dependent (BOLD) activity in both ventral and dorsal striatum during learning tasks patterned after those used to elicit dopaminergic responses in animals (Delgado et al. 2000; McClure et al. 2003; Schönberg et al. 2007; D'Ardenne et al. 2008). Furthermore, imaging studies have shown that activity in structures that modulate dopamine output (caudal orbital frontal cortex and ventral striatum) reflects the magnitude of errors on reward-based learning tasks (Knutson and Cooper 2005).

Motor learning is often broadly classified into two categories: sensorimotor adaptation and sequence learning. Although patients with Parkinson's disease (PD) and Huntington's disease (HD) exhibit mild to moderate impairments in kinematic sensorimotor adaptation tasks (cf. Canavan et al. 1990; Laforce and Doyon 2002; Boulet et al. 2005), this finding has generally not been corroborated by the neuroimaging literature. For example, many investigations of kinematic adaptation tasks, such as when subjects make pointing movements while wearing prism lenses or receiving distorted visual feedback of their actions on a computer display, have not reported BG activation (Clower et al. 1996; Inoue et al. 1997; Imamizu et al. 2000). However, several of these studies were not designed to assess early vs. late stages of learning. More recent neuroimaging studies demonstrate activation in the globus pallidus, putamen, and caudate during the early learning phase of sensorimotor adaptation tasks (gain change: Krakauer et al. 2004; feedback rotation: Seidler et al. 2006). In motor sequence learning, increased putamen activity was correlated with performance during the late phases of learning (Seidler et al. 2005). However, it is not clear whether involvement of striatal pathways during these motor learning tasks is related to error processing because the pathways subserve other functions and the two types of learning rely on differing processes.

In the following section, we outline feedforward and feedback models of basal ganglia contributions to error processing during motor tasks. For more complex computational models of the basal ganglia focusing on action selection and the interactions between multiple corticostriatal circuits, we refer readers to a recent issue of *Current Opinion in Neurobiology* (e.g., Frank 2011; Ito and Doya 2011; and van der Meer and Redish 2011).

Feedforward Error Processing in the Basal Ganglia

Schmidt and Lee (2011) defines two ways that a person can make an error in achieving a task goal. One is an error in motor planning and the other is an error in motor execution. Motor planning errors involve selection of a motor 'program' that is inappropriate for the given situation. Correcting a planning error requires perception of the error and selection of a new action plan. Striatal neurons appear to represent predictive information related to movement and reward, and hence could participate in comparing motor output to an internal model or predicted motor outcomes (Guigon et al. 1995). Schmidt estimated that errors in planning require corrections that have approximately 200 ms latency, because attention is required for correcting an error in selection. The involvement of the BG in such movement planning has been indirectly implicated by studies in humans with BG dysfunction. For example, Smith et al. (2000) report that reaching movements of HD patients begin to become irregular 200–300 ms after movement onset, potentially reflecting a deficit in correcting planning errors. Because corrective actions based on visual (Wolpert et al. 1995) and proprioceptive (Cordo and Flanders 1990) information acquired during reaching movements began to take place at about the time at which HD patients' movements become irregular, the authors suggest that the system that generates these corrective actions is disturbed. In other words, the movement errors seen in HD patients by Smith et al. (2000) might be part of a more general deficit in action planning and selection as opposed to errors in execution or incorporating task feedback.

Further supporting feedforward theories of BG error processing in movement, the dopamine system broadcasts a prediction error signal for reinforcement learning (Schultz 1998; Rangel et al. 2008). Notably, DA neurons in the animal midbrain respond phasically to primary rewards and stimuli that come, via learning, to predict reward (Fig. 3.2). The pattern of these phasic responses resembles a reward prediction error signal derived from formal reinforcement learning models (Dayan and Balleine 2002; Bayer and Glimcher 2005; Daw and Doya 2006; Morris et al. 2006). Neurons in the BG have been shown to predict reward by firing vigorously in advance of reward upon completion of the requirements for reward attainment (Schultz 1998), hinting that predictive capacity may be a general feature of some basal ganglia structures.

Work by Izawa and Shadmehr (2011) suggests that reward prediction error is a significant component of motor adaptation. The authors suggest that the purpose of learning is not merely to estimate the magnitude of a perturbation but to produce motor commands that maximize reward. Based on their theory, an optimal learner utilizes both reinforcement learning for action selection, and state estimation for identifying the sensory consequences of motor commands (Izawa and Shadmehr 2011). Additionally, we (Seidler et al. 2006) have suggested that the left caudate and globus pallidus may contribute to action selection during sensorimotor adaptation tasks as well.

Error signals are relatively preserved in the ventral striatum of early stage PD patients, but impaired in the dorsolateral striatum relative to healthy controls. This pattern reflects the known selective pattern of dopaminergic denervation in PD (Schonberg et al. 2010). Schonberg et al. (2010) showed that prediction error activity in the human striatum of PD patients was differentially affected by disease and was detectably abnormal only in the dorsal putamen, which is innervated by the depleted nigrostriatal pathway. These findings suggest that prediction error signals measured in the human striatum by the BOLD signal likely reflect midbrain phasic DA activity. These results also provide evidence for a deficiency in predictive error signaling in the dorsolateral striatum of PD patients, which may offer an explanation for the deficits observed by these patients in other reward learning tasks. Evidence for an overlapping prediction error signal during learning with juice and money rewards has also been found in a region of dorsal striatum (the caudate nucleus), while prediction error signals in a subregion of ventral striatum were significantly stronger during learning with money but not juice reward

(Valentin and O'Doherty 2009). Taken together, these findings fill in a missing link in the puzzle of the role of prediction error signals in reward-related learning and thus provide additional support to the reinforcement-learning hypothesis of dopamine.

Feedback Error Processing in the Basal Ganglia

Feedback error processing can result in corrective and adaptive actions after an (response) error has occurred. Examples of immediate corrective actions include attempts to inhibit the error online, immediate error corrections, and response time slowing for trials following an error (Rabbitt 1990). The main output of the basal ganglia modulates the action of the thalamus, which relays sensory information to the cortex and basal ganglia (McFarland and Haber 2000). This information stream, Smith et al. (2000) suggest, is likely to participate in error feedback control.

The notion of an error-correction dysfunction following basal ganglia damage is not a new one even though the cerebellum has most often been attributed to error correction functions (Flament et al. 1996). For example, Butters and Rosvold (1968) proposed that the caudate nucleus forms part of a neural mechanism for achieving error correction in the motor system, and Angel et al. (1971) attributed some of the motor deficits observed in PD to slowed error correction mechanisms. In a single unit recording study of basal ganglia activity, in which animals learned a motor sequencing task, cells in the caudate fired only following an incorrect button press, supporting a role for the caudate in mechanisms of error correction.

The idea that the BG nuclei may be critical for signaling online motor decision-making processes—particularly during feedback rather than feedforward-based motor control—fits well with a long-held view of their role in filtering competing motor programs—by inhibiting unwanted motor programs and disinhibiting desired actions (Houk 2005; Houk et al. 2007; Mink 1996; Mink and Thach 1993). The primary output of the BG is inhibitory (via tonically active GABAergic neurons in BG output structures). Thus, the striatum can disinhibit cortical targets by releasing tonic inhibition and selectively boosting activation of the most salient channel. Therefore the BG do not select the actions themselves but rather facilitate their execution via the 'direct pathway' from striatum to BG output structures (Mink 1996). For example, Chevrier and Schachar (2010) show that error detection deactivates the dorsal substantia nigra, dorsal striatum and ventral bank of the ACC. This network is in keeping with observations of phasic suppression of dopamine neurons on error trials (DeLong 1983; Fiorillo et al. 2003; Ljungberg et al. 1991), and the hypothesized role of this pathway in modulating error processes (Amiez et al. 2005; Brown and Braver 2005; Takenouchi et al. 1999). Furthermore, error trials that lead to the greatest response slowing deactivate structures that modulate dopamine output (Eblen and Graybiel 1995) and encode error magnitude (Menon et al. 2007; Murray et al. 2008). For example, the magnitude of ventral striatum deactivation correlates with error magnitude (Murray et al. 2008; O'Doherty et al. 2003).

Gehring et al. (2000) found evidence for exaggerated compensatory behavior in obsessive-compulsive disorder (OCD), and they also concluded that the BG, which might be overactive in OCD, implements action corrections. Such mechanisms also appear relevant to the complex stereotyped behaviors associated with increased dopamine-mediated activity in the striatum (Canales and Graybiel 2000). Cools (1985) applied control systems theory to the results of experimental manipulations of the basal ganglia in animals, which implicates the striatum in arbitrarily programming the ordering and sequencing of behavioral states of varying complexity. On the basis of pharmacological manipulations, Van den Bercken and Cools (1982) concluded that dopamine-mediated activity in the striatum increases the magnitude of error signals, leading to stereotyped behavior.

As mentioned earlier, patients with basal ganglia disorders show little or no deficits in motor adaptation paradigms like force field (Smith and Shadmehr 2005) or visuomotor perturbations (Agostino et al. 1996; Gabrieli et al. 1997). In the typical force field or visuomotor tasks sensory feedback is available, and therefore sensory prediction errors likely play a dominant role in the adaptive process (Izawa and Shadmehr 2011). Learning from sensory prediction errors appears to depend on the integrity of the cerebellum (Synofzik et al. 2008; Tseng et al. 2007), thus the ability of basal ganglia patients to adapt to visuomotor and force field perturbations may provide evidence that changes in motor output in these tasks are primarily driven by sensory prediction errors. Tunik et al. (2009) suggest that the striatum and cerebellum play complementary roles in regulating ongoing actions when precise updating is required. Similar to the symptomatic HD patients, the individuals with cerebellar damage have poorer reaching movement performance than controls, but the decrement in their performance when perturbations are given is generally more like that of controls than HD subjects (Smith et al. 2000). This suggests that subjects with HD generally have greater deficits in error feedback control than do cerebellar patients. Thus the basal ganglia may play a key role in deciding if and when to correct a given movement by initiating corrective submovements, whereas the cerebellum is likely more involved in amplifying and refining the command signals to specify movements with different amplitudes, velocities, and directions.

Cerebellar Contributions to Error Processing

It has long been known that the cerebellum (Fig. 3.3) is important for motor control and learning, based on accounts of patients with lesions to the structure (Holmes 1939). Early theories of cerebellar function proposed that it plays a key role in the learning of new skills (Marr 1969). More specifically, the cerebellum is known to play a role in error-based motor learning (Criscimagna-Hemminger et al. 2010; Ito 2002, Ramnani 2006; Tseng et al. 2007; Miall and Wolpert 1996) and is thought to be important for the formation of internal models of actions (Imamizu et al. 2000; Diedrichsen et al. 2005; for reviews see Ramnani 2006; Ito 2008). The cerebellum may serve as a monitor of motor errors (Marr 1969; Kitazawa et al. 1998; Doya

Fig. 3.3 The *left* panel cartoon illustrates a motor error (art by Lauren Wu); both feedback and feedforward processing of such errors have been linked to the cerebellum (*right* panel)

2000; Desmurget and Grafton 2000; Seidler et al. 2004), and through this process it is involved in the online tuning of movements. It may also use error information to learn and form internal models of action representations.

Most notably, the cerebellum is thought to be an important structure for detecting errors in movement by comparing the predicted with the actual sensorimotor consequences of a movement (Miall et al. 1993; Blakemore et al. 2001; reviewed in Ramnani 2006). Differences between the predicted feedback and the actual feedback are then compared and movements are adjusted accordingly online. Particularly in adaptation learning, the cerebellum is important for detecting errors, but also using those errors to update forward models for online movement control (Kitazawa et al. 1998; Imamizu et al. 2000; Tseng et al. 2007; Izawa and Shadmehr 2011). It is through feedback processes that the cerebellum is able to contribute to feedforward models.

Feedback Error Processing in the Cerebellum

Online correction of movements is thought to be reliant upon the cerebellum. The cerebellum is thought to predict the sensory consequences of action with an efferent copy of the motor commands (Blakemore et al. 2001; Miall et al. 1993). By comparing the actual sensory feedback with the predicted sensory consequences, the cerebellum can detect errors and send corrective signals to the motor cortex to allow for smooth and accurate movements. By detecting and correcting errors during the performance of a task, the cerebellum may aid in learning by updating the existing internal models of a particular motor command.

Adaptation of reaching movements seems to be highly dependent upon sensory prediction errors (Izawa and Shadmehr 2011; Tseng et al. 2007). Patients with hereditary cerebellar ataxia are particularly impaired in sensorimotor adaptation (Tseng et al. 2007). Performance on a reach adaptation task was compared between cerebellar ataxia patients and controls. In one condition the participants were allowed to make motor corrections, while in the other condition participants were unable to correct motor errors. Control participants performed equally well with and without online motor corrections. This indicates that sensory prediction errors are key components for motor adaptation, as the addition of online motor corrections provided no additional gains. Similarly, there was also no difference with and without motor corrections in the cerebellar patient group; however, this group generally showed great deficits in adaptation when compared with controls. Taken together, this indicates that adaptation learning is through sensory prediction errors.

The cerebellar-dependent updating of internal models using sensory feedback seems to largely be an implicit process. Mazzoni and Krakauer (2006) investigated implicit learning and explicit strategies in visuomotor adaptation by giving individuals a specific strategy to employ during task performance. They found that although individuals eliminated errors almost immediately when given explicit information, over the course of adaptation there was an increase in performance error (endpoint location of the cursor). They proposed that this was due to ongoing implicit adaptive processes overriding the explicit strategy. To further investigate and dissociate implicit and explicit processes in visuomotor adaptation learning, Taylor et al. (2010) tested individuals with cerebellar ataxia in a similar paradigm. While both patients and controls reduced errors very quickly when given the explicit strategy, the control group showed the expected increase in endpoint error as training progressed whereas the cerebellar group did not. Thus, the inability of cerebellar patients to implicitly learn allowed for good performance due to the explicit strategy, demonstrating that the cerebellum is critical for implicit adaptive learning.

In a related study, Criscimanga-Hemminger et al. (2010) also demonstrated that sensory prediction errors are key to learning reach adaptation and dependent upon the cerebellum. However, they also found that the size of the error is important. Individuals with cerebellar degeneration were unable to learn from large errors resulting from sudden introduction of a perturbation. However, when the perturbation was gradually increased across trials, learning did occur. They suggest that perhaps other brain regions, or more spared regions of the cerebellum, are involved in learning from small errors, which is why learning occurred. Thus, there seems to be a distinction between learning from small versus large errors.

Functional neuroimaging has also provided insight into feedback error processes in the cerebellum. Grafton et al. (2008) investigated the neural correlates of visuomotor tracking and looked at feedback processes using neuroimaging coupled with models of movement. Feedback responses were those with a delay of 150 ms and due to a response to the position or speed of the cursor. Activity in the cerebellum was strongly correlated with both the magnitude of tracking errors and motor corrections. The authors speculate that error processing in the cerebellum may be related to feedback control while not being specific to the formation of an internal model (Grafton et al. 2008).

To investigate feedback control, we manipulated target size in a joystick aiming task (Seidler et al. 2004). Similarly, Ogawa et al. (2006) manipulated the availability of visual feedback during a visual tracing task. In the case of Seidler et al. (2004), small targets required feedback control. In the visual tracing task (Ogawa et al. 2006), when visual information was available, feedback mechanisms would be employed. The cerebellum was found to be more active under feedback control conditions in both studies. Specifically, we (Seidler et al. 2004) noted a decrease in cerebellar activity when movements were made to larger targets, and Ogawa et al. (2006) found greater activity in the cerebellum when visual information was available. Taken together, these studies support a role for the cerebellum in feedback processes of motor control.

Further supporting this notion is work looking at complex spiking patterns in the cerebellar Purkinje cells of monkeys performing reaching movements Kitazawa et al. (1998) noted a spike in firing that occurred at the end of the arm movement. They proposed that this spike is important for encoding errors allowing for learning, supporting the notion that the cerebellum plays a role in feedback processes.

As briefly described above, feedback from errors may be particularly important for driving the formation of internal models (Kawato and Gomi 1992; Kitazawa et al. 1998; Imamizu et al. 2000). During the early stages of learning in both visuomotor adaptation tasks (Imamizu et al. 2000) and sequence learning (Jueptner et al. 1997; Doyon et al. 1997; Doyon et al. 2003), the fMRI activation is seen in the cerebellum. Imamizu et al. (2000) noted widespread cerebellar activation early on in learning that was proportional to motor errors, but after learning, smaller areas of activation in posterior regions of the cerebellum remained, perhaps due to the acquisition of a new internal model.

Feedforward Error Processing in the Cerebellum

The motor system relies at least in part on forward models to perform and learn smooth coordinated movements (Miall and Wolpert 1996). The learning of internal models for feedforward monitoring of motor performance seems to rely at least in part on the cerebellum (Miall et al. 1993; Miall and Wolpert 1996; Kitazawa et al. 1998; Imamizu et al. 2000; Doya 2000; Lang and Bastian 2002; Morton and Bastian 2004). Computational theories of motor control first suggested that the cerebellum might be the brain structure where feedforward information is formed into internal models for motor control (Miall et al. 1993; Kawato and Gomi 1992). The cytoarchitecture of the cerebellum is such that it could support a supervised learning system wherein motor commands serve as input, and incoming sensory information (errors) act as a ‘teacher’ to then refine the motor commands, and create internal models allowing for feedforward control (Marr 1969; Doya 2000). Both patient and neuroimaging studies support such a role for the cerebellum in feedforward control of movement.

As described above, Kitazawa et al. (1998) recorded Purkinje cell firing while monkeys made reaching movements. In addition to the complex spiking seen at the end of the movement, there was also complex spiking at the beginning of the movement. The authors suggest that this spiking may represent the destination of the reach to aid in feedforward control. Thus, the cerebellum also seems to play a key role in feedforward processes.

Investigations of force field adaptation in cerebellar patients have revealed that patients with cerebellar damage or degeneration do not adapt (Smith and Shadmehr 2005), and also do not generalize their performance to similar though unpracticed tasks (Smith and Shadmehr 2005; Maschke et al. 2004; Morton and Bastian 2004). Smith and Shadmehr demonstrated that although cerebellar patients modified their performance using within trial corrections, they were unable to use their errors to learn (across trial corrections). Furthermore, they were unable to generalize the task when moving in the opposite direction. Maschke et al. (2004) also demonstrated that cerebellar patients show no evidence of learning in a forcefield adaptation task, and again, found that patients were unable to generalize to unpracticed regions of the workspace. Taken together, it is clear that cerebellar patients have adaptation deficits likely due to their inability to process errors across trials, which allows for the creation and updation of internal models.

Lang and Bastian (2002) tested cerebellar patients as well as healthy controls drawing a figure-of-eight with their arms (in a standing position), under single and dual-task conditions. During the figure-of-eight task, participants were told to optimize performance by increasing the number of eights that were traced in the air. While cerebellar patients did show some improvement on the figure of eight task, they had marked difficulty under dual-task conditions and returned to initial performance levels, while control participants showed no interference. Cerebellar patients were unable to automatically perform the task relying upon feedforward mechanisms, and instead needed to rely on cognitive control.

Diedrichsen et al. (2005) provide further evidence for the role of the cerebellum in learning and forming internal models of reaching. They investigated both target and execution errors in a forcefield adaptation task using functional neuroimaging. Target errors are the errors due to unpredictability in the location of a target, whereas execution errors are the result of a miscalibration of internal models. Activity in cerebellar hemispheres V, VI, and VII was seen during execution errors, supporting that this region is important for feedforward control.

Finally, noninvasive brain stimulation of the cerebellum during target aiming movements also provides support for the role of the cerebellum in feedforward movement control (Miall et al. 2007). Transcranial magnetic stimulation (TMS) was administered to lateral regions of the cerebellum during the reach movement to disrupt its processing. Individuals receiving TMS showed greater errors in their movements relative to those that did not receive stimulation. Interestingly, the movement errors were consistent with out-of-date movement estimations from earlier in the movement. The cerebellum seems thus to be important for estimating the state of the arm during reach, likely through feedforward processes.

Caveats

Numerous studies have demonstrated an important role for error experience in motor learning. Reduction of errors via physical guidance has been shown to hinder learning (Domingo and Ferris 2009) and error augmentation has been shown to aid learning processes (Domingo and Ferris 2010; Wei et al. 2005; Patton et al. 2006; Emken and Reinkensmeyer 2005). However, motor learning is not critically dependent on error detection and correction. For example, Diedrichsen et al. (2010) have reported evidence for use-dependent learning, or improvements via repetition of correct movements. Moreover, Wachter et al. (2009) have shown that procedural learning is enhanced with positive reinforcement. Thus, although error detection and correction processes can be important for motor learning, other mechanisms can be relied upon as well. In addition, although it is self-evident, we wish to emphasize that error based learning can only take place in the presence of errors. That is, the small performance adjustments that take place late in the learning process are likely to rely on differing mechanisms. Finally, although error-based learning can be important to early learning, other cognitive processes such as working memory also appear to make substantial, and perhaps related, contributions (for recent reviews see Seidler et al. under review, in press).

Conclusions and Future Directions

There is substantial evidence favoring a role for the cerebellum, medial prefrontal/ACC, and basal ganglia systems in error corrective processes occurring during motor skill learning. Cerebellar networks have long been implicated in these behaviors. Meanwhile, although traditionally viewed as detecting discrete, binary (present or absent), 'cognitive' errors, evidence is accumulating to support a role for medial prefrontal/ACC error processing mechanisms in motor skill learning, including reports that this region scales its activity in a continuous fashion with motor error magnitude. Additionally, data supports a role for basal ganglia pathways in both *within trial* error corrections and reinforcement-dependent error learning mechanisms. What remains to be resolved is whether these pathways act independently or cooperatively during motor learning, and whether such interactions might vary depending on the task to be learned and the environmental context. Recent studies have reported both structural (Bostan et al. 2010; Bostan and Strick 2010) and functional (Kwak et al. 2010) connectivity between the cerebellum and basal ganglia, supporting the plausibility of multiple interactive error correction systems for motor learning.

References

Agostino R, Sanes JN, Hallett M (1996) Motor skill learning in Parkinson's disease. J Neurol Sci 139(2):218–226

Amiez C, Joseph JP, Procyk E (2005) Anterior cingulate error-related activity is modulated by predicted reward. Eur J Neurosci 21(12):3447–3452

Angel RW, Alston W, Garland H (1971) L-dopa and error correction time in Parkinson's disease. Neurology 21(12):1255–1260
Anguera JA, Seidler RD, Ghering WJ (2009) Changes in performance monitoring during sensorimotor adaptation. J Neurophys 102:1868–1879
Anguera JA, Reuter-Lorenz PA, Willingham DT, Seidler RD (2010) Contributions of spatial working memory to visuomotor learning. J Cogn Neurosci 22(9):1917–1930
Badgaiyan RD, Posner MI (1998) Mapping the cingulate cortex in response selection and monitoring. Neuroimage 7(3):255–260
Bayer HM, Glimcher PW (2005) Midbrain dopamine neurons encode a quantitative reward prediction error signal. Neuron 47(1):129–141
Berns GS, Cohen JD, Mintun MA (1997) Brain regions responsive to novelty in the absence of awareness. Science 276(5316):1272–1275
Blakemore S-J, Frith CD, Wolpert DM (2001) The cerebellum is involved in predicting the sensory consequences of action. Neuro Report 12:1879–1884
Bostan AC, Dum RP, Strick PL (2010) The basal ganglia communicate with the cerebellum. Proc Natl Acad Sci 107(18):8452–8456
Bostan AC, Strick PL (2010) The cerebellum and basal ganglia are interconnected. Neuropsychol Rev 20(3):261–270
Botvinick MM, Braver TS, Barch DM, Carter CS, Cohen JD (2001) Conflict monitoring and cognitive control. Psychol Rev 108(3):624–652
Boulet C, Lemay M, Bédard MA, Chouinard MJ, Chouinard S, Richer F (2005) Early Huntington's disease affects movements in transformed sensorimotor mappings. Brain Cogn 57(3):236–243
Brown JW, Braver TS (2005) Learned predictions of error likelihood in the anterior cingulate cortex. Science 307(5712):1118–1121
Butters N, Rosvold HE (1968) Effect of caudate and septal nuclei lesions on resistance to extinction and delayed-alternation. J Comp Physiol Psychol 65(3):397–403
Canales JJ, Graybiel AM (2000) A measure of striatal function predicts motor stereotypy. Nat Neurosci 3(4):377–383
Canavan AG, Passingham RE, Marsden CD, Quinn N, Wyke M, Polkey CE (1990) Prism adaptation and other tasks involving spatial abilities in patients with Parkinson's disease, patients with frontal lobe lesions and patients with unilateral temporal lobectomies. Neuropsychologia 28(9):969–984
Chao HH, Luo X, Chang JL, Li CS (2009) Activation of the pre-supplementary motor area but not inferior prefrontal cortex in association with short stop signal reaction time—an intra-subject analysis. BMC Neurosci 10:75
Chen LL, Wise SP (1996) Evolution of directional preferences in the supplementary eye field during acquisition of conditional oculomotor associations. J Neurosci 16(9):3067–3081
Chen X, Scangos KW, Stuphorn V (2010) Supplementary motor area exerts proactive and reactive control of arm movements. J Neurosci 30(44):14657–14675
Chevrier A, Schachar RJ (2010) Error detection in the stop signal task. Neuroimage 53(2):664–673
Clower DM, Hoffman JM, Votaw JR, Faber TL, Woods RP, Alexander GE (1996) Role of posterior parietal cortex in the recalibration of visually guided reaching. Nature 383(6601):618–621
Cools AR (1985) Morphine and specific changes in the sensitivity of noradrenergic receptors within the 'limbic' part of the feline caudate nucleus: a behaviour study. Brain Res Bull 14(3):239–250
Cordo PJ, Flanders M (1990) Time-dependent effects of kinesthetic input. J Mot Behav 22(1):45–65
Criscimanga-Hemminger SE, Bastian AJ, Shadmehr R (2010) Size of error affects cerebellar contributions to motor learning. J Neurophysiol 103:2275–2284
Cunnington R, Windischberger C, Deecke L, Moser E (2002) The preparation and execution of self-initiated and externally triggered movement: a study of event-related fMRI. Neuroimage 15(2):373–385

Danckert J, Ferber S, Goodale MA (2008) Direct effects of prismatic lenses on visuomotor control: an event-related functional MRI study. Eur J Neurosci 28(8):1696–1704
D'Ardenne K, McClure SM, Nystrom LE, Cohen JD (2008) BOLD responses reflecting dopaminergic signals in the human ventral segmental area. Science 319(5867):1264–1267
Daw ND, Doya K (2006) The computational neurobiology of learning and reward. Curr Opin Neurobiol 16(2):199–204
Dayan P, Balleine BW (2002) Reward, motivation, and reinforcement learning. Neuron 36(2):285–298
Dehaene S, Posner MI, Tucker DM (1994) Localization of a neural system for error-detection and compensation. Psychol Sci 5(5):303–305
Delgado A, Sierra A, Querejeta E, Valdiosera RF, Aceves J (2000) Inhibitory control of the GABAergic transmission in the rat neostriatum by D2 dopamine receptors. Neuroscience 95(4):1043–1048
DeLong MR (1983) The neurophysiologic basis of abnormal movements in basal ganglia disorders. Neurobehav Toxicol Teratol 5(6):611–616
den Ouden HE, Daunizeau J, Roiser J, Friston KJ, Stephan KE (2010) Striatal prediction error modulates cortical coupling. J Neurosci 30:3210–3219
Desmurget M, Grafton S (2000) Forward modeling allows feedback control for fast reaching movements. Trends Cogn Sci 4:423–431
Desmurget M, Vindras P, Grea H, Viviani P, Grafton ST (2000) Proprioception does not quickly drift during visual occlusion. Exp Brain Res 134(3):363–377
Dhar M, Pourtois G (2011) Early error detection is generic, but subsequent adaption to errors is not: evidence from ERPs. Neuropsychologia 49(5):512–517
Diedrichsen J, Hashambhoy Y, Rane T, Shadmehr R (2005) Neural correlates of reach errors. J Neurosci 25:9919–9931
Diedrichsen J, White O, Newman D, Lally N (2010) Use-dependent and error-based learning of motor behaviors. J Neurosci 30(15):5159–5166
Domingo A, Ferris DP (2009) Effects of physical guidance on short-term learning of walking on a narrow beam. Gait Posture 30(4):464–468
Domingo A, Ferris DP (2010) The effects of error augmentation on learning to walk on a narrow balance beam. Exp Brain Res 206(4):359–370
Donchin O, Francis JT, Shadmehr R (2003) Quantifying generalization from trial-by-trial behavior of adaptive systems that learn with basis functions: theory and experiments in human motor control. J Neurosci 23(27):9032–9045
Doya K (2000) Complementary roles of basal ganglia and cerebellum in learning and motor control. Curr Opin Neurobiol 10:732–739
Doyon J, Gaudreau D, Laforce Jr R, Castonguay M, Bedard PJ, Bedard F, Bouchard J-P (1997) Role of the striatum, cerebellum, and frontal lobes in the learning of a visuomotor sequence. Brain Cogn 34:218–245
Doyon J, Penhune V, Ungerleider LG (2003) Distinct contribution of the cortico-striatal and cortico-cerebellar systems to motor skill learning. Neuropsychologia 41:252–262
Duann JR, Ide JS, Luo X, Li CS (2009) Functional connectivity delineates distinct roles of the inferior frontal cortex and presupplementary motor area in stop signal inhibition. J Neurosci 29(32):10171–10179
Eblen F, Graybiel AM (1995) Highly restricted origin of prefrontal cortical inputs to striosomes in the macaque monkey. J Neurosci 15(9):5999–6013
Eimer M, Goschke T, Schlaghecken F, Sturmer B (1996) Explicit and implicit learning of event sequences: evidence from event-related brain potentials. J Exp Psychol Learn Mem Cogn 22(4):970–987
Emken JL, Reinkensmeyer DJ (2005) Robot-enhanced motor learning: accelerating internal model formation during locomotion by transient dynamic amplification. IEEE Trans Neural Syst Rehabil Eng 13:33–39
Falkenstein M, Hohnsbein J, Hoormann J (1995) Event-related potential correlates of errors in reaction tasks. Electroencephalogr Clin Neurophysiol Suppl 44:287–296

Ferdinand NK, Mecklinger A, Kray J (2008) Error and deviance processing in implicit and explicit sequence learning. J Cogn Neurosci 20(4):629–642
Fiorillo CD, Tobler PN, Schultz W (2003) Discrete coding of reward probability and uncertainty by dopamine neurons. Science 299(5614):1898–1902
Flament D, Ellermann JM, Kim SG, Ugurbil K, Ebner TJ (1996) Functional magnetic resonance imaging of cerebellar activation during the learning of a visuomotor dissociation task. Hum Brain Mapp 4(3):210–226
Flanagan JR, Vetter P, Johansson RS, Wolpert DM (2003) Prediction precedes control in motor learning. Curr Biology 13(2):146–150
Folstein JR, Van Petten C (2008) Influence of cognitive control and mismatch on the N2 component of the ERP: a review. Psychophysiology 45(1):152–170
Frank MJ (2011) Computational models of motivated action selection in corticostriatal circuits. Curr Opin Neurobiol 21(3):381–386
Gabrieli JD, Stebbins GT, Singh J, Willingham DB, Goetz CG (1997) Intact mirror-tracing and impaired rotary-pursuit skill learning in patients with Huntington's disease: evidence for dissociable memory systems in skill learning. Neuropsychology 11(2):272–281
Gehring WJ, Goss B, Coles MGH, Meyer DE, Donchin E (1993) A Neural System for Error-Detection and Compensation. Psychol Sci 4(6):385–390
Gehring WJ, Coles MG, Meyer DE, Donchin E (1995) A brain potential manifestation of error-related processing. Electroencephalogr Clin Neurophysiol Suppl 44:261–272
Gehring WJ, Himle J, Nisenson LG (2000) Action-monitoring dysfunction in obsessive-compulsive disorder. Psychol Sci 11(1):1–6
Gehring WJ, Liu Y, Orr JM, Carp J (2011) The error-related negativity (ERN/Ne) In: Luck SJ, Kappenman E (eds) Oxford handbook of event-related potential components. Oxford University Press, New York
Goldman-Rakic PS, Leranth C, Williams SM, Mons N, Geffard M (1989) Dopamine synaptic complex with pyramidal neurons in primate cerebral cortex. Proc Natl Acad Sci USA 86(22):9015–9019
Grafton ST, Schmitt P, Van Horn J, Diedrichsen J (2008) Neural substrates of visuomotor learning based on improved feedback control and prediction. NeuroImage 39:1383–1395
Guigon E, Dorizzi B, Burnod Y, Schultz W (1995) Neural correlates of learning in the prefrontal cortex of the monkey: a predictive model. Cereb Cortex 5(2):135–147
Hebb DO (1949) The organization of behavior. Wiley, New York
Hikosaka O, Sakai K, Miyauchi S, Takino R, Sasaki Y, Putz B (1996) Activation of human presupplementary motor area in learning of sequential procedures: a functional MRI study. J Neurophysiol 76(1):617–621
Hochman EY, Eviatar Z, Breznitz Z, Nevat M, Shaul S (2009) Source localization of error negativity: additional source for corrected errors. Neuroreport 20(13):1144–1148
Holmes G (1939) The cerebellum of man. Brain 62:1–30
Holroyd CB, Coles MG (2002) The neural basis of human error processing: reinforcement learning, dopamine, and the error-related negativity. Psychol Rev 109(4):679–709
Horvitz JC (2002) Dopamine gating of glutamatergic sensorimotor and incentive motivational input signals to the striatum. Behav Brain Res 137(1–2):65–74
Houk JC (2005) Agents of the mind. Biol Cybern 92(6):427–437
Houk JC, Bastianen C, Fansler D, Fishbach A, Fraser D, Reber PJ et al (2007) Action selection and refinement in subcortical loops through basal ganglia and cerebellum. Philos Trans R Soc Lond B Biol Sci 362(1485):1573–1583
Imamizu H, Miyauchi S, Tamada T, Sasaki Y, Takino R, Pütz B, Yoshioka T, Kawato M (2000) Human cerebellar activity reflecting an acquired internal model of a new tool. Nature 403:192–195
Inoue K, Kawashima R, Satoh K, Kinomura S, Goto R, Sugiura M et al (1997) Activity in the parietal area during visuomotor learning with optical rotation. Neuroreport 8(18):3979–3983
Isoda M, Hikosaka O (2007) Switching from automatic to controlled action by monkey medial frontal cortex. Nat Neurosci 10(2):240–248

Ito M (2002) Historical review of the significance of the cerebellum and the role of purkinje cells in motor learning. Ann N Y Acad Sci 978:273–288
Ito M (2008) Control of mental activities by internal models in the cerebellum. Nat Rev Neurosci 9:304–313
Ito M, Doya K (2011) Multiple representations and algorithms for reinforcement learning in the cortico-basal ganglia circuit. Curr Opin Neurobiol, 21(3):368–373
Izawa J, Shadmehr R (2011) Learning from sensory and reward prediction errors during motor adaptation. PLoS Comput Biol 7(3):e1002012
Jueptner M, Frith CD, Brooks DJ, Frackowiak RSJ, Passingham RE (1997) Anatomy of motor learning. II. Subcortical structures and learning by trial and error. J Neurophysiol 77:1325–1337
Kawato M, Gomi H (1992) A computational model of four regions of the cerebellum based on feedback-error learning. Biol Cybern 68:95–103
Kitazawa S, Kimura T, Yin P-B (1998) Cerebellar complex spikes encode both destinations and errors in arm movements. Nature 392:494–497
Knutson B, Cooper JC (2005) Functional magnetic resonance imaging of reward prediction. Curr Opin Neurol 18(4):411–417
Krakauer JW, Ghilardi MF, Mentis M, Barnes A, Veytsman M, Eidelberg D et al (2004) Differential cortical and subcortical activations in learning rotations and gains for reaching: a PET study. J Neurophysiol 91(2):924–933
Krigolson OE, Holroyd CB (2006) Evidence for hierarchical error processing in the human brain. Neuroscience 137(1):13–17
Krigolson OE, Holroyd CB (2007a) Hierarchical error processing: different errors, different systems. Brain Res 1155:70–80
Krigolson OE, Holroyd CB (2007b) Predictive information and error processing: the role of medial-frontal cortex during motor control. Psychophysiology 44(4):586–595
Krigolson OE, Holroyd CB, Van Gyn G, Heath M (2008) Electroencephalographic correlates of target and outcome errors. Exp Brain Res 190(4):401-411
Kwak Y, Peltier SJ, Muller MLTM, Bohnen N, Dayalu P, Seidler RD (2010) Altered resting state cortico-striatal connectivity in mild to moderate stage Parkinson's disease. Front Sys Neurosci 4:143
Laforce R, Doyon J (2002) Differential role for the striatum and cerebellum in response to novel movements using a motor learning paradigm. Neuropsychologia 40(5):512–517
Lang CE, Bastian AJ (2002) Cerbellar damage impairs automaticity of a recently practiced movement. J Neurophysiol 87:1336–1347
Ljungberg T, Apicella P, Schultz W (1991) Responses of monkey midbrain dopamine neurons during delayed alternation performance. Brain Res 567(2):337–341
Logan G (1994) On the ability to inhibit thought and action. In: Dagenbach D, Carr TH (eds) Inhibitory Processes in attention. Memory and language. Academic Press, San Diego
Marr D (1969) A theory of cerebellar cortex. J Physiol 202:437–470
Maschke M, Gomez CM, Ebner TJ, Konczak J (2004) Hereditary cerebellar ataxia progressively impairs force adaptation during goal-directed arm movements. J Neurophysiol 91:230–238
Mazzoni P, Krakauer JW (2006) An implicit plan overrides an explicit strategy during visuomotor adaptation. J Neurosci 26:3642–3645
McClure SM, Berns GS, Montague PR (2003) Temporal prediction errors in a passive learning task activate human striatum. Neuron 38(2):339–346
McFarland NR, Haber SN (2000) Convergent inputs from thalamic motor nuclei and frontal cortical areas to the dorsal striatum in the primate. J Neurosci 20(10):3798–3813
Menon M, Jensen J, Vitcu I, Graff-Guerrero A, Crawley A, Smith MA et al (2007) Temporal difference modeling of the blood-oxygen level dependent response during aversive conditioning in humans: effects of dopaminergic modulation. Biol Psychiatry 62(7):765–772
Miall RC, Weir DJ, Wolpert DM, Stein JF (1993) Is the cerebellum a smith predictor? J Mot Behav 25:203–216

Miall RC, Wolpert DM (1996) Forward models for physiological motor control. Neural Netw 9:1265–1279
Miall RC, Christensen LOD, Cain O, Stanley J (2007) Disruption of state estimation in the human lateral cerebellum. PLoS Biol 5:e316
Miltner WHR, Braun CH, Coles MG H (1997) Event-related brain potentials following incorrect feedback in a time-estimation task: evidence for a 'generic' neural system for error detection. J Cogn Neurosci 9(6):788–798
Mink JW, Thach WT (1993) Basal ganglia intrinsic circuits and their role in behavior. Curr Opin Neurobiol 3(6):950–957
Mink JW (1996) The basal ganglia: focused selection and inhibition of competing motor programs. Prog Neurobiol 50(4):381–425
Morris SE, Yee CM, Nuechterlein KH (2006) Electrophysiological analysis of error monitoring in schizophrenia. J Abnorm Psychol 115(2):239–250
Morton SM, Bastian AJ (2004) Prism adaptation during walking generalizes to reaching and requires the cerebellum. J Neurophysiol 92:2497–2509
Murray GK, Corlett PR, Clark L, Pessiglione M, Blackwell AD, Honey G et al (2008) Substantia nigra/ventral tegmental reward prediction error disruption in psychosis. Mol Psychiatry 13(3):239, 267–276
Nachev P, Rees G, Parton A, Kennard C, Husain M (2005) Volition and conflict in human medial frontal cortex. Curr Biol 15(2):122–128
Nachev P, Kennard C, Husain M (2008) Functional role of the supplementary and pre-supplementary motor areas. Nat Rev Neurosci 9(11):856–869
O'Doherty JP, Dayan P, Friston K, Critchley H, Dolan RJ (2003) Temporal difference models and reward-related learning in the human brain. Neuron 38(2):329–337
Ogawa K, Inui T, Sugio T (2006) Separating brain regions involved in internally guided and visual feedback control of moving effectors: an event-related fMRI study. NeuroImage 32:1760–1770
Patton JL, Stoykov ME, Kovic M, Mussa-Ivaldi FA (2006) Evaluation of robotic training forces that either enhance or reduce error in chronic hemiparetic stroke survivors. Exp Brain Res 168:368–383
Prodoehl J, Yu H, Wasson P, Corcos DM, Vaillancourt DE (2008) Effects of visual and auditory feedback on sensorimotor circuits in the basal ganglia. J Neurophysiol 99(6):3042–3051
Rabbitt P (1990) Age, IQ and awareness, and recall of errors. Ergonomics 33(10-11):1291–1305
Rabbitt PM (1966) Errors and error correction in choice-response tasks. J Exp Psychol 71(2):264–272
Ramnani N (2006) The primate cortico-cerebellar system: anatomy and function. Nat Rev Neurosci 7:511–522
Rangel A, Camerer C, Montague PR (2008) A framework for studying the neurobiology of value-based decision making. Nat Rev Neurosci 9(7):545–556
Ridderinkhof KR, Ullsperger M, Crone EA, Nieuwenhuis S (2004) The role of the medial frontal cortex in cognitive control. Science 306(5695):443–447
Russeler J, Kuhlicke D, Munte TF (2003) Human error monitoring during implicit and explicit learning of a sensorimotor sequence. Neurosci Res 47(2):233–240
Schmidt RA, Lee TD (2011) Motor control and learning: a behavioral emphasis. Human Kinetics, Champaign
Schönberg T, Daw ND, Joel D, O'Doherty JP (2007) Reinforcement learning signals in the human striatum distinguish learners from nonlearners during reward-based decision making. J Neurosci 27(47):12860–12867
Schonberg T, O'Doherty JP, Joel D, Inzelberg R, Segev Y, Daw ND (2010) Selective impairment of prediction error signaling in human dorsolateral but not ventral striatum in Parkinson's disease patients: evidence from a model-based fMRI study. Neuroimage 49(1):772–781
Schultz W (1998) Predictive reward signal of dopamine neurons. J Neurophysiol 80(1):1–27
Schultz W (2001) Reward signaling by dopamine neurons. Neuroscientist 7(4):293–302

Seidler RD, Noll DC, Thiers G (2004) Feedforward and feedback processes in motor control. NeuroImage 22:1775–1783
Seidler RD, Purushotham A, Kim SG, Ugurbil K, Willingham D, Ashe J (2005) Neural correlates of encoding and expression in implicit sequence learning. Exp Brain Res 165(1):114–124
Seidler RD, Noll DC, Chintalapati P (2006) Bilateral basal ganglia activation associated with sensorimotor adaptation. Exp Brain Res 175(3):544–555
Seidler, RD, Bo J, Anguera JA (under review). Neurocognitive contributions to motor skill learning: the role of working memory. Invited paper for a special issue of the Journal of Motor Behavior dedicated to papers from the Neural Control of Movement Society's 2011 satellite meeting on motor learning.
Seidler RD, Benson BL, Boyden NB, Kwak Y (in press) Motor skill learning. In: Ochsner KN, Kosslyn SM (eds), Oxford handbook of cognitive neuroscience. Oxford University Press, New York
Shadmehr R, Mussa-Ivaldi F (1994) Adaptive representation of dynamics during learning of a motor task. J Neurosci 14:3208–3224
Shadmehr R, Smith MA, Krakauer JW (2010) Error correction, sensory prediction, and adaptation in motor control. Annu Rev Neurosci 33:89–108
Shima K, Mushiake H, Saito N, Tanji J (1996) Role for cells in the presupplementary motor area in updating motor plans. Proc Natl Acad Sci USA 93(16):8694–8698
Smith AD, Bolam JP (1990) The neural network of the basal ganglia as revealed by the study of synaptic connections of identified neurones. Trends Neurosci 13(7):259–265
Smith MA, Brandt J, Shadmehr R (2000) Motor disorder in Huntington's disease begins as a dysfunction in error feedback control. Nature 403(6769):544–549
Smith MA, Shadmehr R (2005) Intact ability to learn internal models of arm dynamics in Huntington's disease but not cerebellar degeneration. J Neurophysiol 93(5):2809–2821
Spraker MB, Yu H, Corcos DM, Vaillancourt DE (2007) Role of individual basal ganglia nuclei in force amplitude generation. J Neurophysiol 98(2):821–834
Stuphorn V, Brown JW, Schall JD (2010) Role of supplementary eye field in saccade initiation: executive, not direct, control. J Neurophysiol 103(2):801–816
Synofzik M, Lindner A, Thier P (2008) The cerebellum updates predictions about the visual consequences of one's behavior. Curr Biol 18(11):814–818
Takenouchi K, Nishijo H, Uwano T, Tamura R, Takigawa M, Ono T (1999) Emotional and behavioral correlates of the anterior cingulate cortex during associative learning in rats. Neuroscience 93(4):1271–1287
Taylor JA, Klemfuss NM, Ivry RB (2010) An explicit strategy prevails when the cerebellum fails to compute movement errors. Cerebellum 9:580–586
Tseng Y, Diedrichsen J, Krakauer JW, Shadmehr R, Bastian AJ (2007) Sensory prediction errors drive cerebellum-dependent adaptation of reaching. J Neurophysiol 98:54–62
Tunik E, Houk JC, Grafton ST (2009) Basal ganglia contribution to the initiation of corrective submovements. Neuroimage 47(4):1757–1766
Turner RS, Grafton ST, Votaw JR, Delong MR, Hoffman JM (1998) Motor subcircuits mediating the control of movement velocity: a PET study. J Neurophysiol 80(4):2162–2176
Turner RS, Desmurget M, Grethe J, Crutcher MD, Grafton ST (2003) Motor subcircuits mediating the control of movement extent and speed. J Neurophysiol 90(6):3958–3966
Vaillancourt DE, Mayka MA, Thulborn KR, Corcos DM (2004) Subthalamic nucleus and internal globus pallidus scale with the rate of change of force production in humans. Neuroimage 23(1):175–186
Valentin VV, O'Doherty JP (2009) Overlapping prediction errors in dorsal striatum during instrumental learning with juice and money reward in the human brain. J Neurophysiol 102(6):3384–3391
Van den Bercken JH, Cools AR (1982) Evidence for a role of the caudate nucleus in the sequential organization of behavior. Behav Brain Res 4(4):319–327

van der Meer MA, Redish AD (2011) Ventral striatum: a critical look at models of learning and evaluation. Curr Opin Neurobiol 21(3):387–392

Wachter T, Lungu OV, Liu T, Willingham DT, Ashe J (2009) Differential effect of reward and punishment on procedural learning. J Neurosci 29(2):436–443

Wei Y, Bajaj P, Scheidt R, Patton JL (2005) Visual error augmentation for enhancing motor learning and rehabilitative relearning. In: International conference on rehabilitation robotics. IEEE, Chicago, pp 505–510

Wolpert DM, Ghahramani Z, Jordan MI (1995) Are arm trajectories planned in kinematic or dynamic coordinates? An adaptation study. Exp Brain Res 103(3):460–470

Wolpert DM, Miall RC, Kawato M (1998) Internal models in the cerebellum. Trends Cogn Sci 2:338–347

Chapter 4
Plasticity in the Motor Network Following Primary Motor Cortex Lesion

Numa Dancause

Topographic and Anatomical Organization of the Primary Motor Cortex

Original brain mapping studies using surface stimulation revealed the basic pattern of somatostopic representation in the primary motor cortex (M1), where the leg is represented most medially followed by the trunk, arm, neck, face, and mouth as one moves laterally along the precentral gyrus (Penfield and Boldrey 1937; Phillips and Porter 1977; Woolsey et al. 1952). These studies also reported an orderly somatotopy within the arm representation, represented in both the famous homunculus and simiusculus. In subsequent studies, the general location of body parts along the central sulcus essentially remained unchanged. However, within the forelimb area, extensive evidence accumulated suggesting that instead of being somatotopically organized, cortical areas controlling movements of digits, wrist, and forearm (pronation and supination) are intermingled with one another. For example, using intracortical microstimulation techniques (ICMS) in squirrel monkeys, we consistently find a mosaic of distal forelimb (DFL) movements bordered by proximal representation of the arm, except caudally, where nonresponsive cortex corresponding to the somatosensory area is found (Fig. 4.1). A similar "horseshoe-like" organization of the arm representation has also been reported in macaque monkeys using stimulus-triggered averaging of electromyographic activity (Griffin et al. 2009; Park et al. 2001).

The absence of a detailed somatotopy within the hand representation was clearly shown in a study of single-neuron activity in M1 of macaques (Schieber and Hibbard 1993). In this study, neurons were recorded while monkeys performed isolated flexion or extension of individual fingers and of the wrist in different trials. Results indicated

N. Dancause (✉)
Groupe de Recherche sur le Système Nerveux Central (GRSNC), Département de Physiologie, Pavillon Paul-G-Desmarais, Université de Montréal, 2960, Chemin de la Tour, bureau 4138, Montréal, Québec H3T 1J4, Canada
e-mail: numa.dancause@umontreal.ca

M. J. Richardson et al. (eds.), *Progress in Motor Control,*
Advances in Experimental Medicine and Biology 782,
DOI 10.1007/978-1-4614-5465-6_4,

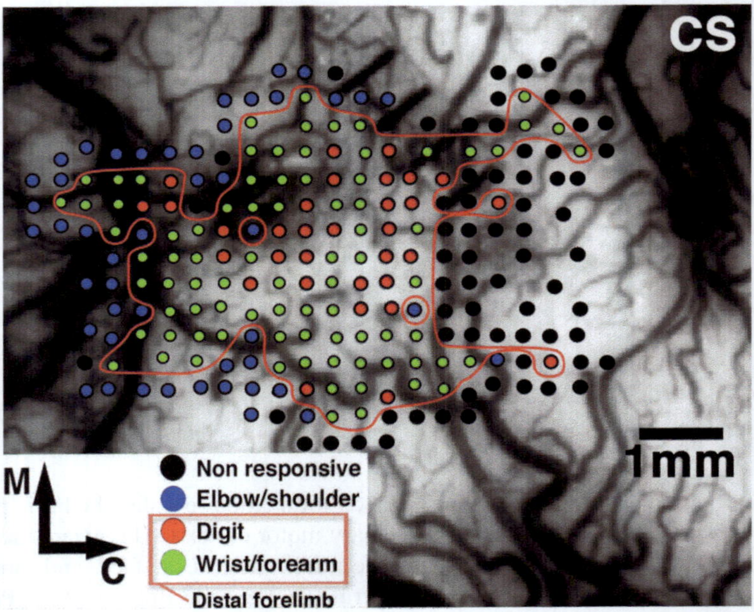

Fig. 4.1 Typical topographic map of the primary motor cortex in squirrel monkeys. Organization of the motor map was explored with intracortical microstimulation methods. Each *dot* on the map corresponds to a location where the electrode was lowered at depth corresponding to layer 5 and microcurrent ($<30\ \mu A$) was applied. Interpenetration distance was approximately 250 μm. Movements evoked are grouped in three color-coded categories. A *red* contour surrounds the DFL movements, which consist of intermingled digits *(red)*, wrist *(green)*, and forearm *(green)* movements. The DFL movements are bordered by proximal *(blue)* movements medially, rostrally, and laterally. Caudally, nonresponsive sites *(black)* are found in areas that correspond to the primary somatosensory cortex. *C* caudal, *CS* central sulcus, and *M* medial

that neurons with activity related to the movements of each finger were not clustered within one specific cortical region. Rather, they were scattered throughout a large cortical area and were intermingled with neurons related to the movement of other fingers and of the wrist. Interestingly, the absence of clear somatotopy in M1 is also reflected in the thalamocortical inputs to M1. Neurons in the ventrolateral thalamic nucleus (VL) can be antidromically stimulated from widely separated locations in M1 (Shinoda et al. 1985), supporting that projections of thalamocortical neurons to M1 are also diverging to extended and discontinuous cortical areas.

The mosaic organization within the DFL representation of M1 places neurons involved in the control of different muscles of the forearm and hand close to one another. This proximity could favor rapid integration of neural signals across the entire arm representation to favor the coordination of different segments during various arm movements (Schieber 2001). In light of the preceding literature, it may be more appropriate to refer to the conglomerate of digit, wrist, and forearm movements as the DFL representation rather than the *hand representation* per se. Accordingly, the term DFL will be used in the following sections.

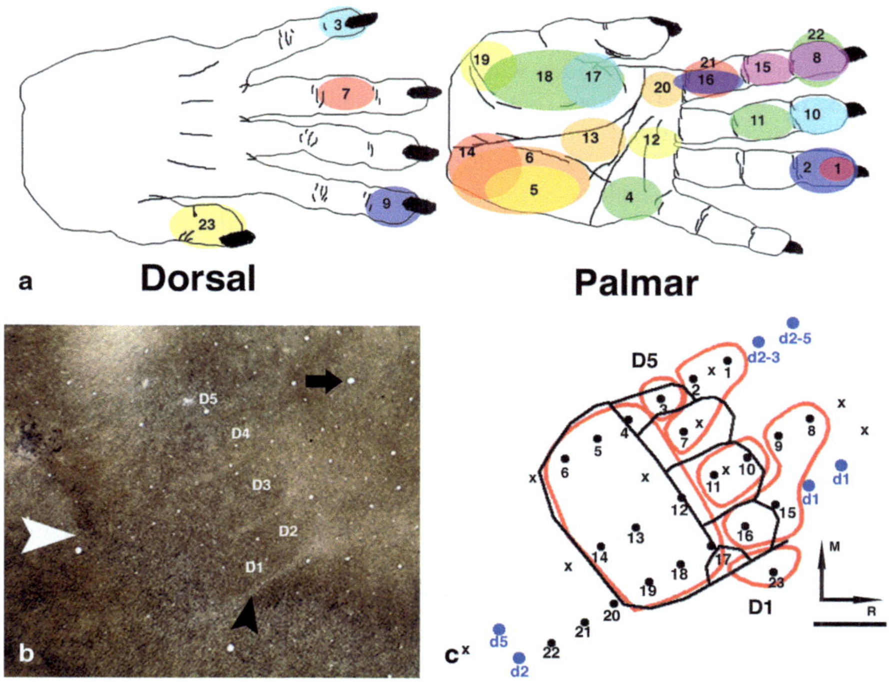

Fig. 4.2 Area 3b hand representation defined by multiunit recording of cutaneous receptive fields and myeloarchitecture. **a** Orderly arrangement of receptive fields in S1. Numbers in the receptive fields correspond to the numbers on the physiological map shown in **c**. Note the somatotopic reversal between sites 19 and 20, indicating the border between area 3b and area 1/2. **b** Area 3b hand representation shown with myelin staining in a tangential section through the ipsilateral cortex. *White arrowhead* shows the caudal border of the hand area. *Black arrowhead* shows the hand–face septum. *Black arrow* shows an example of a blood vessel used for the alignment of myelin section with sensory map. Brightness and contrast were adjusted. *M* medial, *R* rostral, and scale bar 1 mm. **c** Superposition of the myeloarchitectonically defined hand area (*black contour*) onto sensory maps obtained with multiunit recordings used to define the area 3b hand representation (*red contour*) in the same case. Alignment of blood vessel locations was used to coregister the two sets of data. Each recording site is represented by a *dot*. Sites where no response was elicited are indicated by *x*. Sites where responses were evoked by joint movements, but not by cutaneous stimulation, are indicated by *blue dots*. Receptive field is indicated beneath each *blue dot*. Scale bar 1 mm. (Adapted from Dancause et al. 2006b)

The apparent absence of topographic organization in the DFL representation of M1 contrasts with that of the primary somatosensory cortex (S1). In this area, digit representations are typically found in predictable locations, with receptive fields of digit 5 located medially and receptive fields of digits 4, 3, 2, and 1 progressively more laterally (Kaas 1993). Precise topographic organization is particularly clear in area 3b (Fig. 4.2a). In this subdivision of S1, cutaneous receptive fields are small and the representation of each finger is remarkably well defined and isolated from the representations of other fingers.

The topographic organization within the hand representation of S1 is also supported by anatomical findings. For example, isolation of finger representations in area 3b can be visualized with histochemical staining of myelin in tangential sections (Fig. 4.2b). This technique reveals much denser axonal projections within each finger representation and fewer projections between finger representations. The cortical location of each digit revealed histochemically corresponds very well to the physiological representation of the same digit using multiunit recording of evoked potentials (Fig. 4.2c).

Finally, the corticocortical interconnections between subdivisions of S1 tend to be within consistent topographic representations. For example, injection of a neuroanatomical tracer in the representation of the second digit in area 1 of S1 results in the labeling of neurons located in the representation of the same digit in area 3b (Florence et al. 1998).

In contrast to the anatomical patterns observed in S1, injection of anatomical tracer in a cortical location where stimulation elicits digit movements in M1 results in substantial labeling of the entire forelimb representation (Huntley and Jones 1991). Anatomical divergence is also found at the level of M1 corticospinal outputs. Pyramidal-tract axons can show extensive intraspinal branching, projecting to motoneuron pools of several different muscles (Shinoda et al. 1979, 1986). Spike-triggered averaging reveals that most corticomotoneuronal cells individually project to two to four synergistic muscles (Buys et al. 1986; Cheney and Fetz 1985; Fetz and Cheney 1980; Fetz et al. 1976; Lemon et al. 1987) and that approximately half of the corticospinal cells facilitate movement of at least one proximal and one distal muscle (McKiernan et al. 1998).

Recently, we have also shown that interconnections between M1 and a premotor area, the ventral premotor (PMv) cortex, are not between consistent topographic representations. Instead, outputs from PMv DFL to M1 are topographically divergent (Dancause et al. 2006a). Using ICMS, we identified forelimb representations of M1 and PMv and injected a neuroanatomical tracer in PMv DFL. We then documented the location of synaptic boutons in M1 in relation to the topography and analyzed their distribution in function of the location of the injection core (Fig. 4.3). Whereas the injection core area was predominantly located within PMv DFL (79.0 ± 9.6 %), labeled terminals were found over larger proportions of proximal representations in M1 (40.5 ± 2.8 %). These results suggest that neuronal populations within the PMv DFL send topographically divergent outputs to M1.

When considering the level of anatomical divergence within the motor network, it appears that neurons in M1 or the premotor cortex can excite the motoneuron pools of any muscle of the arm within two synapses. Thus, evoked movements from cortical stimulation in motor areas do not reveal the full extent of anatomical connections between the stimulated neurons. Instead, the emergent properties or the output effects of this efferent population reflect the facilitation of only one path among several preexisting pathways. This facilitated output pathway could be selected through a

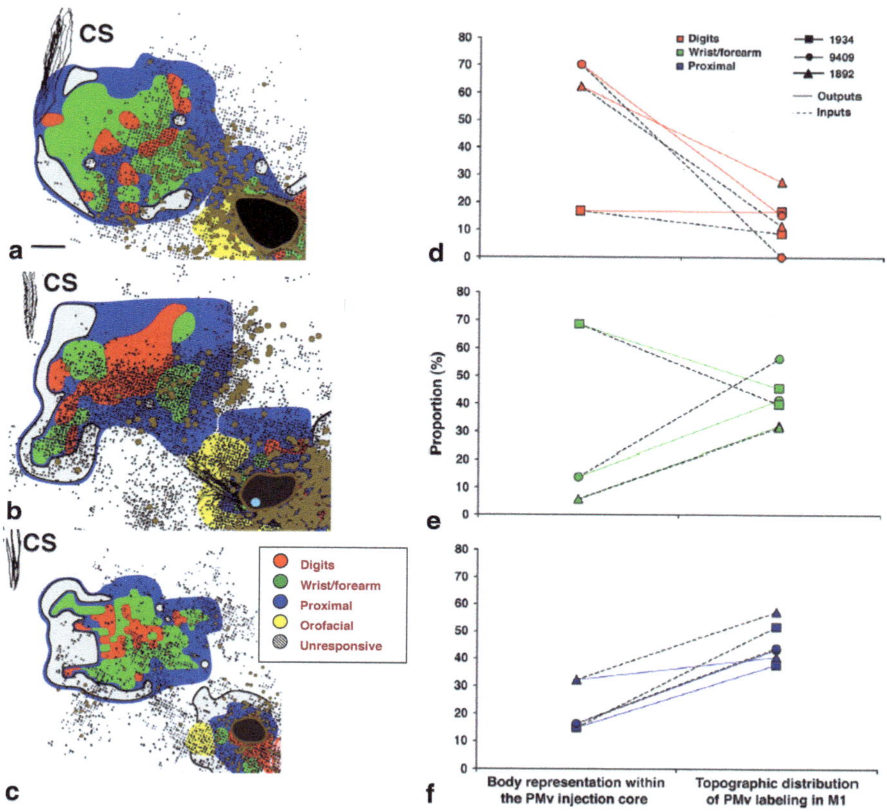

Fig. 4.3 Heterotopic connectivity between PMv and M1. **a–c** Distribution of labeling in the M1 forelimb area following injection of biotinylated dextran amine in PMv in three different monkeys. The small *black dots* show the distribution of labeled terminals and the *large brown dots* show labeled cell bodies. The *black area* surrounded by *brown contour* shows the location of the injection core in PMv. *Left* is caudal and *up* is medial. *CS* central sulcus, scale bar 1 mm. **d** Proportions of digit representation included in the injection core in relation to the proportion of cell bodies and terminals located in the digit representation in M1. **e** Proportions of wrist/forearm representation included in the injection core in relation to the proportion of cell bodies and terminals located in the wrist/forearm representation in M1. **f** Proportions of proximal representation included in the injection core in relation to the proportion of cell bodies and terminals located in the proximal representation in M1. Whereas the injection core was mainly located within DFL territory in PMv, comparable proportions of outputs to M1 and inputs from M1 were located in DFL and proximal representations. *Dashed black line* cell bodies, *full colored line* terminals, *red* digit, *green* wrist/forearm, and *blue* proximal. (Adapted from Dancause et al. 2006a)

competitive process based on behavioral demands, experience, and reward history. If so, this system would be particularly well suited to sustain rapid and dramatic reorganization of corticospinal output effects supporting learning (Nudo et al. 1996b; Plautz et al. 2000) or recovery after injury (Nudo et al. 1996c).

Movement Representations in Topographic Maps Correlate with Function

As discussed in previous sections, ICMS can be used to produce topographic maps of motor areas. Repeated production of these maps within one animal at different time points was first used to explore cortical plasticity within M1 of normal (uninjured) monkeys and provided a basic description of fundamental properties of cortical plasticity relative to the distal and proximal forelimb representation in M1 (Nudo et al. 1996b). In squirrel monkeys, topographic maps derived with ICMS in the same animal at different time points with no intervening manipulation reveal remarkably constant proportions of digit, wrist, forearm, elbow, and shoulder territories. However, when monkeys are trained to retrieve pellets from a small well, a task that requires them to develop a novel movement strategy with the digits, the cortical area from which digit movements are evoked increases. If the monkeys are then trained on a different task requiring a novel strategy involving wrist movements, the cortical territory devoted to digit movements is reduced to favor expansion of the representation of wrist movements. In comparison, monkeys simply performing repetitive movements of a natural task, which does not require learning of a novel strategy, show similar proportions of digit, wrist, forearm, elbow, and shoulder territories as do naive animals (Plautz et al. 2000). Functional reorganization of motor maps is thus dependent on prior behavioral experience. That is, the cortical area from which one can evoke a particular movement correlates with the use of this specific movement in behavioral tasks, if learning is involved.

Comparable cortical reorganization with behavioral learning has also been shown in S1. For example, when monkeys are trained on a frequency-discrimination task in which a vibratory stimulus is applied at a constant location on a digit (Recanzone et al. 1992b), the increase in performance is associated with an increase of the cortical area devoted to the digit involved in the task (Recanzone et al. 1992a, c). In contrast, passive stimulation of the digits has little effect on the cortical representational maps (Recanzone et al. 1992b); once again suggesting that learning is a key component for the devotion of additional cortical area to the representation of a body segment.

Together, these results support the idea that there is a competition for the limited cortical "real estate" dictated by behavioral requirements. Learning of novel skills would require a greater cortical contribution, which can be progressively lessened as the skills required by the task are acquired. The freed cortical area could then serve in the learning of other skills.

Primary Motor Cortex Lesion Models to Study Recovery of Hand Function

Stroke is a complicated, multifactorial affliction. Damage in most patients is of variable size and location, and may disrupt gray matter, white matter, and subcortical structures. Following stroke, there is often disturbance of motor function of the arm and leg. Even well-recovered patients still suffer from chronic motor deficits, particularly in regard to hand function (Lai et al. 2002). Thus, understanding the mechanisms that support recovery of hand control is an especially important issue in neurorehabilitation.

The role of M1 in the control of arm and hand movements has been extensively studied (Ebner et al. 2009; Georgopoulos and Stefanis 2007; Glees and Cole 1950; Kalaska 2009; Kalaska et al. 1998; Schieber 2001; Woolsey et al. 1952). These studies have shown that M1 is involved in the fine control of force, movement and posture of the contralateral distal arm. M1 is the largest source of corticospinal neurons (Dum and Strick 1991, 2002; He et al. 1993). Many of these corticospinal projections reach lower cervical segments where motoneurons controlling the hand are located (Dum and Strick 1991, 2002; He et al. 1993, 1995). Moreover, in some primate species, these projections target the ventral horn in the spinal cord gray matter, suggesting that they have monosynaptic connections onto motoneurons (Ralston and Ralston 1985). The monosynaptic contacts between cortical projections and motoneurons have recently been confirmed with transneuronal labeling using retrograde transport of rabies virus (Rathelot and Strick 2006, 2009). In these experiments, virus was injected in muscles of the arm and incubated to allow transport across one synapse. The injections resulted in labeling of M1 cells, mainly in its caudal portion.

The corticomotoneuronal system of M1 is believed to be crucial for the control of complex and fractionated finger movements (Bennett and Lemon 1996; Porter 1985). A major argument in favor of this hypothesis is that species with projections from M1 to the anterior horn of the spinal cord are typically more dexterous. A good example comes from the comparison of corticospinal projection patterns of two New World primates sharing the same environment, the capucin and squirrel monkey. The capucin is capable of fractionated finger movements and opposition of the thumb and has numerous corticospinal projections to the ventral horn. In contrast, the squirrel monkey is only capable of prehensile grasps, consisting of simultaneous flexions of digits, and has very sparse, if any, projections to the ventral horn (Bortoff and Strick 1993; Maier et al. 1997).

Lesion and inactivation studies in animal models also support the importance of M1 and its corticospinal projections for the control of hand movements. Lesions or inactivation affecting M1 have dramatic detrimental effects on motor function (Fogassi et al. 2001; Fulton 1935; Glees and Cole 1950; Kubota 1996; Nudo and Milliken 1996a). These studies show that disturbing M1 results in loss of dexterity and movement speed, muscle weakness, and increased reaction time. Even inactivating very small regions in the M1 DFL with injections of a GABA agonist produces weakness, slowness, and loss of independence of given finger movements (Schieber and Poliakov 1998). This contrasts dramatically with the effects of a lesion or inactivation of premotor areas, which result in subtle consequences to the function of the hand (Brinkman 1984; Kermadi et al. 1997; Kurata and Hoffman 1994; Rizzolatti et al. 1983; Schieber 2000). For example, lesion or inactivation in PMv typically produces reluctance in using the contralateral hand (Rizzolatti et al. 1983; Schieber 2000) and inactivation of supplementary motor area (SMA) produces initiation delays in bimanual trials (Kermadi et al. 1997), both without causing obvious motor deficits with that hand once the movements are initiated.

In human studies, whereas the relationship between lesion location and functional outcomes has not been as straightforward, the disruption of corticospinal projections appears to better correlate with motor function than lesion size itself (Pineiro et al. 2000; Zhu et al. 2010). Cortical lesions including M1 are a frequent consequence of middle cerebral artery occlusion, the most common source of strokes. In humans, the destruction of M1 has particularly devastating effects on motor function and greatly contributes to motor deficits of the hand after stroke. To study the mechanisms supporting the recovery of hand function, experimenters have taken advantage of animal models to control for the effects of lesion size and location in M1 and have investigated the plasticity associated with functional recovery.

M1 Lesion Recovery and Perilesional Plasticity

In animal models, it is possible to identify the M1 DFL using ICMS and then to induce lesions in specific regions within this area. In squirrel monkeys, even targeted lesions as small as 3–3.5 mm^2, which corresponds to approximately 30 % of the total cortical area devoted to the M1 DFL, cause clear deficit of the contralateral hand function (Nudo and Milliken 1996a). Postlesion monitoring of motor recovery with a modified Klüver board, from which monkeys retrieve small food pellets from wells of different diameters, shows that monkeys rely more on the hand ipsilateral to the lesion and are less efficient in retrieving pellets with the hemiparetic hand. However, different patterns of deficit are observed depending on whether the lesion affects the caudal or the rostral portion of the DFL (Friel et al. 2005). After a lesion to rostral M1, animals often make aiming errors. Instead of reaching directly into the well to grasp the food pellet, they often touch the surface of the board outside the well before entering the well. After a lesion to caudal M1, they typically reach directly into the well but have to visually confirm the presence of the pellet in their hands. The differences in deficits may be related to the pattern of sensory inputs to M1 and the sensory information processed by M1 neurons. Indeed, neurons with cutaneous receptive fields are more common in the caudal portion of M1 and neurons with proprioceptive receptive fields are more common in the rostral portion of M1 (Strick and Preston 1982). Deficits resulting from a lesion in caudal M1 may be due to problems in the processing of cutaneous information or the failure to integrate cutaneous information with motor commands. In contrast, deficits following lesion in rostral M1 would be due to a dysfunction in the processing of proprioceptive information or the failure to integrate proprioceptive information with motor commands (Friel et al. 2005).

Mapping of the topographic organization with ICMS before and after recovery has been used to investigate the plasticity associated with recovery from a lesion. The strength of this approach is that reorganization associated with recovery can be analyzed within each animal instead of comparing control and experimental groups, as is necessary in human studies. After a small cortical lesion in the caudal portion of the M1 DFL, if a monkey does not receive behavioral training and mapping is not redone 3 months after the lesion, a further reduction of the DFL area in the perilesional

M1 can be observed. In these animals, stimulation of sites that evoked digit and wrist movements in the perilesional M1 prior to lesion are either nonresponsive or rather evoke shoulder and elbow movements (Nudo and Milliken 1996a). This loss of the DFL representation in spontaneously recovered animals could suggest that recovery was associated with suboptimal cortical plasticity, in which the surviving tissue in M1 was not, or was insufficiently, involved in recovery.

If monkeys undergo a rehabilitative treatment, consisting of restricting the use of the unimpaired limb with a jacket and repetitive training of the impaired arm on a retrieval task, the area devoted to M1 DLF is preserved and can even increase after the recovery period. This rehabilitative treatment has the same effect on cortical reorganization on the perilesional cortex if lesions affect the caudal or rostral portion of M1 (Friel et al. 2007).

It thus appears that reorganization of the surviving M1 tissue can support recovery from small cortical infarcts and that this mechanism is maximized with rehabilitative treatment. Similarly in humans after stroke, it seems that the reorganization of activation patterns involving the surviving M1 is associated with better recovery. Whereas poorly recovered patients are more likely to show increased activation of various areas during movements, well-recovered patients show patterns of activation that are comparable to healthy controls (Carey et al. 2006). However, in some patients the lesion is too extensive, leaving insufficient or no tissue in M1.

Recruiting Plasticity in Distant Areas to Support Recovery

Following a small lesion, the spared tissue within the affected area can reorganize and support the recovery of the lost function (Lashley 1929, 1930). In the case of large lesions, when damage is too extensive or the entire area is destroyed, the affected area is insufficient to recover the functional loss. In these cases, recovery must call upon the reorganization of other distant areas that may not have been originally involved in the function more directly mediated by the injured zone, but that can undergo adaptive plasticity to vicariously assume the lost function (Glees and Cole 1950; Munk 1881; Ogden and Franz 1917).

Following lesion in M1, the premotor areas are particularly well positioned to take over the lost M1 function. Indeed, these areas share extensive interconnections with other areas within the motor network, they have their own corticospinal outputs, and already boast motor-related activity prior to the lesion.

Premotor Areas in Nonhuman Primates

In addition to M1, several premotor areas are closely involved in the production of motor outputs. Premotor areas are defined as frontal areas that have direct access to M1 and to the spinal cord (Dum and Strick 2002; Fulton 1935). Six premotor areas have been identified. Most medially, the rostral, dorsal, and ventral cingulate areas (CMAr, CMAd, and CMAv, respectively) are located in the cingulate sulcus (Morecraft and Van Hoesen 1992; Picard and Strick 1996). Progressively dorsal and lateral are the SMA, dorsal premotor (PMd), and PMv areas (Barbas and Pandya 1987; Matelli et al. 1985, 1991; Morecraft and Van Hoesen 1992; Picard and Strick 1996; Takada et al. 2001; Von Bonin and Bailey 1947). These premotor areas are the main sources of inputs to the rostral part of M1 (Dancause et al. 2006b; Dum and Strick 2002; Rouiller et al. 1994; Stepniewska et al. 1993). Premotor areas are also the origin of a large proportion (~44 %) of corticospinal fibers (Dum and Strick 1991; He et al. 1993, 1995), which terminate in the ventral horn and the intermediate zone in the spinal cord (Ralston and Ralston 1985).

Further subdivisions of several premotor areas into rostral and caudal portions have been proposed. These subdivisions are well documented in macaques (Barbas and Pandya 1987; Matelli et al. 1985; Vogt and Vogt 1919; Von Bonin and Bailey 1947) and have been confirmed by immunohistochemistry (Geyer et al. 2000). For example, PMd and SMA each contain a rostral portion called pre-PMd (F7) and pre-SMA (F6) and a caudal portion called PMd proper (F2) and SMA proper (F3; Picard and Strick 2001). Comparable subdivisions of PMd and SMA have been reported in the New World primates (cebus monkeys (Dum and Strick 2005) and owl monkeys (Preuss et al. 1996; Sakai et al. 2000) as well as in prosimian primates (galagos; Fang et al. 2005).

The caudal subdivisions of PMd and SMA have more connections with M1 and more corticospinal projections than their rostral counterparts (Dum and Strick 1991; He et al. 1993). In contrast, the rostral subdivisions are more interconnected with frontal and prefrontal areas (Bates and Goldman-Rakic 1993; Lu et al. 1994; Luppino et al. 1993). Thalamocortical connectivity of the rostral and caudal subdivisions also differs, suggesting that they receive different sensory information and are part of different subcortical loops (Matelli et al. 1989; Wiesendanger and Wiesendanger 1985). In the rostral subdivision, current intensity to evoke movements with ICMS is generally higher (Luppino et al. 1991; Matelli et al. 1991) and movements tend to be more complex, typically involving multiple joints (Matsuzaka and Tanji 1996; Rizzolatti et al. 1990; Shima et al. 1996). Neuron recording in SMA has shown that the onset of activity in the rostral subdivision generally precedes the activity in the caudal one. For example, changes of neuronal activity can often be observed long before the onset of movement in pre-SMA, whereas activity changes in relation to the actual movement are more frequent in SMA-proper (Alexander and Crutcher 1990; Matsuzaka et al. 1992; Rizzolatti et al. 1990). This suggests that the rostral division of premotor areas is more implicated in the preparatory states of movement production.

As previously discussed, the nature of functional deficits resulting from a lesion or inactivation also suggests that premotor areas are not essential for movement execution per se. Rather, they seem to play a role in movement preparation and adaptation in relation to sensory inputs such as visual information and cognitive or emotional states.

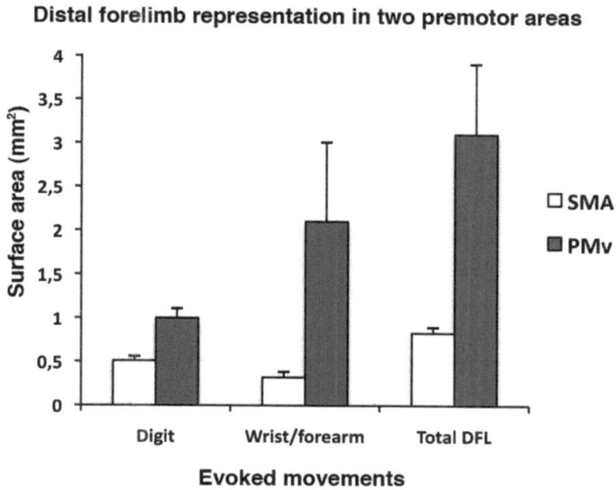

Fig. 4.4 Distal forelimb (DFL) representation surface area in the SMA and PMv cortex in squirrel monkeys. The DFL in the SMA is much smaller than in PMv cortex. (Mean values and standard deviations for SMA was obtained from five monkeys (Eisner-Janowicz et al. 2008) and data for PMv was obtained from four monkeys (Dancause et al. 2008))

Certain Particularities of the Ventral Premotor Cortex

With ICMS techniques, movements of the forelimb, neck, and face can be elicited from PMv. The cortical area devoted to digit movements is large in comparison to other premotor areas. In squirrel monkeys, we systematically find much larger DFL representation in PMv (Dancause et al. 2006c; Frost et al. 2003) than PMd (Dancause et al. 2007) or SMA (Eisner-Janowicz et al. 2008; Fig. 4.4). This is somewhat surprising when considering the corticospinal projections from PMv extend only as far as the upper cervical segments (C2–C4) and projections from other premotor areas are found in upper as well as lower cervical segments (He et al. 1993). Corticospinal outputs of PMv thus only reach segments where motoneurons mainly innervate muscles of the neck and scapula. In the absence of direct corticospinal projections to motoneuron pools of hand muscles, corticospinal neurons of PMv could evoke digit movements oligosynaptically, through the propriospinal cervical network (Alstermark et al. 1990; Pierrot-Deseilligny 1996). However, there is little evidence for such propriospinal transmission in macaque monkeys (Maier et al. 1998). An alternative pathway for PMv to carry its effect on hand muscles is through its projections to M1. Studies conducted in sedated macaques have demonstrated that PMv can have a powerful facilitatory effect on M1 corticospinal outputs (Cerri et al. 2003; Shimazu et al. 2004) and reversible inactivation of M1 reduces or eliminates the capacity to evoke hand movements from PMv (Schmidlin et al. 2008). More recent studies have revealed that the effects of PMv on M1 outputs are complex and task dependent. In awake monkeys, it was shown that the facilitation from PMv is dependent on the object being grasped (Prabhu et al. 2009). In humans, stimulation of PMv at various times during reaching showed that PMv exerts a net inhibitory influence on M1 outputs at rest. During power grip, this inhibition disappears and during precision grip it becomes a net facilitation (Davare et al. 2008). Whereas these data do not rule out that some of the effects on hand muscles observed when stimulating PMv occur at a subcortical site, they do show that PMv has complex and powerful interactions with M1 outputs.

Cytoarchitectonic subdivisions of PMv into caudal and rostral parts have been identified only in macaques so far (Matelli et al. 1985). In fact, rostral PMv (PMvr or area F5) of macaques is characterized by properties that are quite different than those of pre-SMA and pre-PMd. For example, the DFL representation of PMv is found in PMvr (Gentilucci et al. 1988; Rizzolatti et al. 1988) and it has dense connections with M1 (Matelli et al. 1986; Shimazu et al. 2004). It is interesting to note that the DFL representation of PMv in squirrel monkeys (Dancause et al. 2008) and macaque monkeys share many features such as the distance from M1 (Gentilucci et al. 1988), stimulation intensity to evoke movement (Hepp-Reymond et al. 1994), latencies for ICMS-evoked EMG activity (Boudrias and Cheney 2006) and patterns of cortical connections (Barbas and Pandya 1987; Dancause et al. 2006b; Ghosh and Gattera 1995; Matelli et al. 1986). It is thus likely that the PMv DFL representation in squirrel monkeys is the homolog of the DFL representation located in PMvr (or F5) of macaques.

In a recent ICMS study in squirrel monkeys, we identified an additional, isolated DFL representation located rostrally and laterally to the DFL representation of PMv (Dancause et al. 2008). This area, which we generically called the frontal rostral (FR) area, required much higher current intensity to evoke movement and the activity evoked occurred with much longer latencies (Fig. 4.5). It has sparse connections with M1 and numerous connections with the prefrontal cortex and anterior operculum, a pattern that is quite distinct from that of PMv. Anatomical studies in macaques have revealed the existence of a subfield in PMv (area F5a) which shares a similar pattern of cortical connections as the squirrel monkey's FR (Gerbella et al. 2011). Overall, the electrophysiological and anatomical features of FR are very similar to what has been described for pre-PMd and pre-SMA, and it is tempting to propose that FR is the "pre-PMv" of squirrel monkeys. If so, as for pre-SMA and pre-PMd (Picard and Strick 2001), FR should be considered a motor-related field providing an interface between the prefrontal cortex and cortical motor areas, rather than a premotor area per se. In contrast, the PMv DFL representation in squirrel monkeys and PMvr (F5) of macaques would be the equivalent of PMd and SMA proper.

Plasticity in the Ventral Premotor Cortex Following Lesions in M1

The physiological reorganization of PMv associated with recovery from an M1 lesion has been investigated in a series of experiments performed in squirrel monkeys. In each animal, topographic maps of the DFL in M1 and PMv were documented with ICMS. Large ischemic lesions were then induced in M1, destroying more than 50 % of the DFL (Frost et al. 2003). The lesions resulted in an immediate and permanent switch of hand preference. Monkeys were less successful at retrieving pellets with

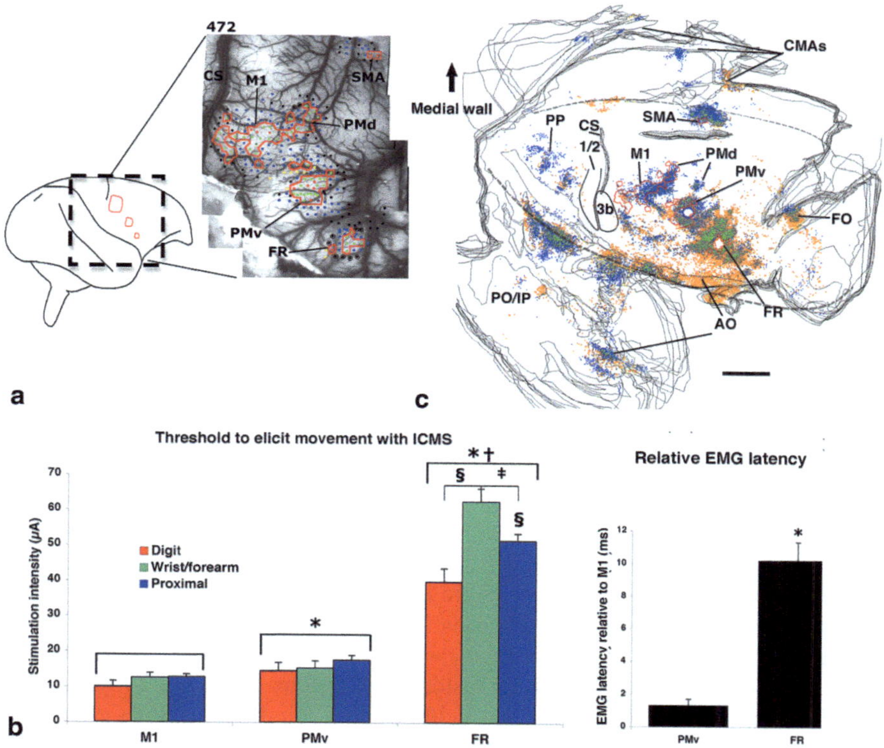

Fig. 4.5 An additional motor-related area frontal and lateral to the ventral premotor cortex. **a** Reconstruction of ICMS movement maps of the lateral frontal cortex in a squirrel monkey (case 472). Digit and wrist/forearm movements comprised the DFL representation, indicated with a *red* contour. Each *dot* represents a microelectrode penetration site. The location of the ICMS-defined motor fields in each case is identified. The cartoon on the *left* shows the approximate location of M1, PMv, and FR. The *dotted square* shows the approximate location of the anatomical reconstruction shown in **c**. *Black dots* nonresponsive, *blue* proximal movements, *green* wrist/forearm, *red* digit, *yellow* orofacial, *R* rostral, *M* medial, and *CS* central sulcus. **b** Thresholds at which movements could be evoked using ICMS in M1, PMv, and FR (mean intensity ± SEM). Statistical analyses revealed that movement thresholds in M1 were significantly lower than PMv and FR (*) and that thresholds in PMv were lower than FR (†). Within FR, digit movement thresholds were significantly lower than proximal and wrist/forearm (§). In addition, wrist/forearm movement thresholds were significantly higher than proximal movement thresholds (‡). **c** Within case comparison of the pattern of connections of PMv and FR DFL representation in the ipsilateral hemisphere. Distribution of labeled terminals following injection of BDA into FR DFL (*orange dots*; see Fig. 4.3) and labeled cell bodies after injection of FB into PMv DFL (*blue dots*). One *orange dot* represents a voxel with labeled terminals and one *blue dot* represents a cell body. When both Fast-Blue and BDA were colocalized, *green* is used. Abbreviations as in Fig. 4.3. Scale bar 5 mm. (Adapted from Dancause et al. 2008)

the hemiparetic hand on the Klüver board, and when successful, they required more finger flexions and required more time in the wells to retrieve the pellets. In the weeks following the injury, monkeys progressively improved their performance with the hemiparetic hand.

After 3 months of spontaneous recovery, ICMS mapping was redone and revealed an increase of PMv cortical surface area devoted to movements of the DFL ranging from 7.2 to 53.8 %. This increase of the PMv DFL representation was proportional to the size of the lesion in M1 and possibly reflects a novel role in the control of the hemiparetic hand.

Is reorganization of distant cortical areas such as PMv really dependent on the quantity of remaining M1 tissue as previously suggested, or could the reorganization be driven by the disruption of the numerous connections between PMv and M1? To address this question, we estimated the potential disruption of cortical connections between PMv and M1 caused by a lesion based on the pattern of connections in control animals and the location of the lesion (Dancause et al. 2006c). In addition to the animals previously described (Frost et al. 2003), we added animals with smaller ischemic lesions (< 50 % of the total M1 DFL representation) located in specific regions of M1 (Fig. 4.6). In two of these animals, lesions were induced in the caudal and medial aspect of the DFL area, where very few connections with PMv are found. In the two other animals, the lesions were located in the rostral and lateral aspect of M1, where the majority of connections with PMv are found. Whereas lesions were of comparable volumes in all four cases, we estimated that the caudomedial lesions destroyed only 6.2 ± 3.0 % and the rostrolateral lesions destroyed 71.7 ± 0.71 % of the interconnections between PMv and M1. Interestingly, following the 3-month recovery period, ICMS mapping revealed that the cortical area devoted to movements of the hand in PMv actually decreased in all four animals. When pooling the data from these animals with the ones from the previous study (Frost et al. 2003), we found that the disruption of M1–PMv connectivity was a poor predictor of the reorganization of the PMv DFL area ($R^2 = 0.3018$; $p = 0.1255$). In contrast, the proportion of the M1 DFL area destroyed by the lesion was a much better predictor ($R^2 = 0.87$; $p = 0.0002$; Fig. 4.4b). The data in Fig. 4.4 also suggest that there is a threshold, approximately corresponding to 50 % of the M1 DFL area, of cortical damage to elicit plastic changes in PMv. Whereas smaller lesions result in a decrease of the PMv DFL area, larger lesions result in reorganization of PMv that is proportional to the lesion size in M1 (Lashley 1929, 1930). It is possible that recovery following small lesions causing relatively mild and transient impairments (Friel et al. 2005) requires only minor postlesion learning (or compensatory motor strategies). In these cases, the intact, adjacent neuronal network in M1 would be sufficient. After larger lesions with severe deficits, the significant motor learning and compensations required cannot be supported by the remaining M1 tissue and may need to call for the vicarious support of distant areas such as PMv.

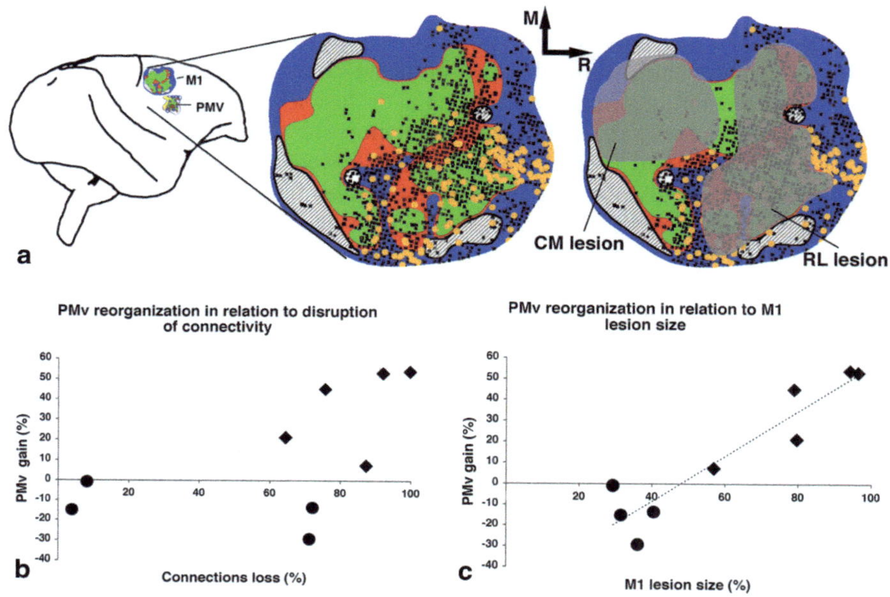

Fig. 4.6 Factors driving PMv reorganization following M1 lesion. **a** Cartoon of a squirrel monkey brain showing the approximate location of the M1 and PMv DFL. The typical distribution of labeled cell bodies (*large orange dots*) and terminals (*small black dots*) in M1 DFL following injection of BDA in PMv in a control case (large injection; case #1934; Dancause et al. 2006b) are shown. Based on these data, $M1_{rl}$ lesions destroy the majority of PMv connections with M1 DFL whereas $M1_{cm}$ lesions have a minor impact. *Red* digit representations, *green* wrist/forearm, *blue* proximal, and *hatched black* nonresponsive. **b** Relationship between the predicted disruption of connectivity of PMv with M1 and physiological reorganization in PMv. Based on cortical connections between PMv and M1 in naive squirrel monkeys, the potential impact of the lesions in M1 were estimated in relation to the physiological reorganization found in PMv. In general, this relationship was not strong. PMv connections were derived as the average of the proportion of cell bodies and terminals. *Circles* are used for the cases with small lesions (Dancause et al. 2006c); *diamonds* are used for cases with larger lesions (Frost et al. 2003). **c** Relationship between PMv reorganization and lesion size in M1. Percent change in PMv DFL 12 weeks postinfarct as a function of the percent anatomical loss of M1 DFL. Whereas animals with larger lesions showed an increase in PMv DFL, animals with smaller lesions showed a decrease. A *diagonal dotted line* shows the linear regression fitting the distribution

Anatomical Rewiring of PMv After M1 Lesions

In another series of experiments, we investigated if sprouting or formation of new polysynaptic connections could account for the reorganization of cortical representational maps in PMv following large lesions in M1. As before, we defined PMv and M1 forelimb representations with ICMS techniques and destroyed a large portion of the M1 DFL representation ($> 50\ \%$) with an ischemic infarct that resulted in an enlargement of the PMv DFL area. At least 5 months after injury, we injected an anatomical tracer in the center of the PMv DFL representation to document its pattern of intracortical connections. Comparison of connections between recovered and experimentally naive animals revealed that following a lesion, PMv axons had atypical orientations and abrupt changes in trajectory at the lesion border (Fig. 4.7a, b). Atypical axons were not only visible in every case at the rostral border of the lesion but the axonal path observed in the experimental group showed clear statistical differences in comparison to the control group. Whereas the average path in control animals assumed a caudomedial orientation, as expected since PMv is located laterally to M1, the path had a caudolateral orientation following recovery from the lesion.

To identify potential area(s) that could share these novel connections with PMv, we analyzed the pattern of connections throughout the entire ipsilesional hemisphere. We found that there was a dramatic increase of connections between PMv and S1 (Dancause et al. 2005), more precisely in area 1/2 of S1 (Fig. 4.7c, d). Whereas these novel connections could be the result of axonal sprouting of multiple preexisting routes, the presence of axons along the rostral–lateral border of the lesion, and their abrupt change in orientation, suggests that sprouting occurred from axons originally terminating in the M1 DFL area that was destroyed. Moreover, looking at the tangential sections at lower magnification revealed the presence of large axons running laterally to the lesion, which appeared to originate from the rostral border of the lesion and were directed toward S1. The increase of neuronal growth-promoting genes in the perilesional cortex (Carmichael et al. 2005) might provide a substrate to favor this pathway. Formation of new connections between PMv and S1 might result in a "bypassing" of the lesion site in M1, perhaps to reestablish a sensorimotor integration loop (Asanuma and Pavlides 1997).

While the techniques we used in these studies do not allow conclusive statements on the precise location of branching or the role of this anatomical reorganization in functional recovery, it is clear that major reorganization of PMv connections occurs as a result of an M1 lesion. Sprouting of new connections between the frontal and parietal cortex should be considered as a potential compensatory mechanism through which the central nervous system accomplishes functional recovery after a lesion in M1.

The changes observed in relation to PMv reviewed in the previous two sections raise the possibility that following M1 lesion any of the premotor areas could undergo substantial physiological and anatomical remodeling during recovery.

Evidence that Other Premotor Areas Can Play a Vicarious Role Following M1 Lesions

In the largely liscencephalic brain of the squirrel monkey, the study of PMv, among premotor areas, has several advantages. The DFL area is entirely located on the cortical surface, facilitating the sampling of electrophysiological data at different times during recovery based on the pattern of surface blood vessels. The proximity of PMv and M1 DFL representations also makes them both accessible within a

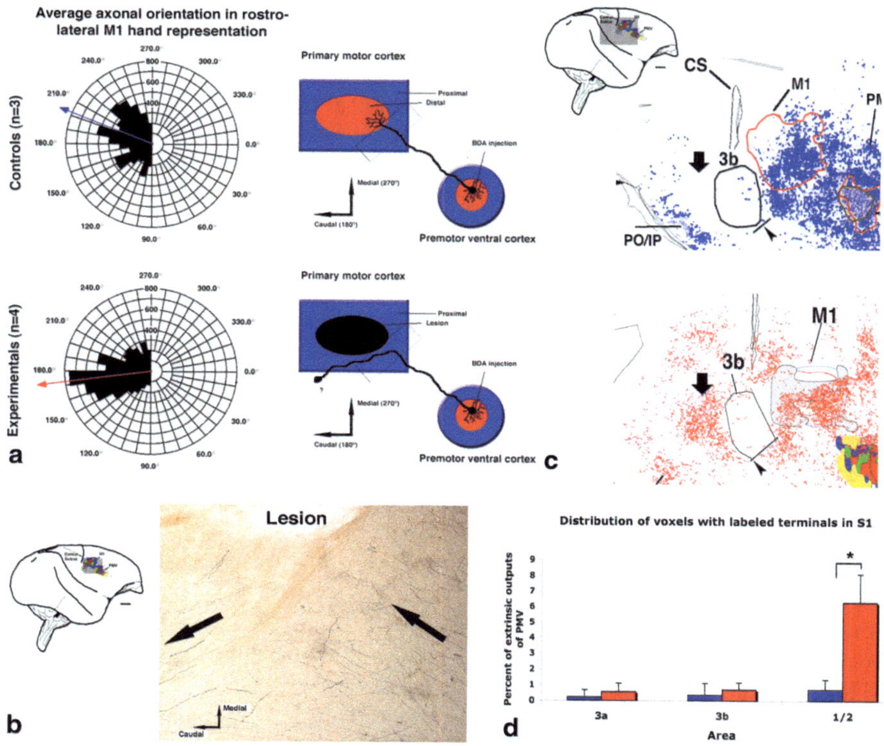

Fig. 4.7 Anatomical rewiring of PMv following M1 lesion. **a** Polar histogram illustrating the distribution of large fiber (axonal) orientations at the rostral border of the M1 lesion. After the M1 injury, fiber orientation is more focused and was directed in a more caudolateral direction. Cartoons on the *left* of the histograms illustrate the presumed alterations in PMv intracortical connections following M1 injury. After PMv targets in M1 are destroyed, PMv intracortical fibers are thought to seek new targets; axons abruptly change orientation near the lesion border, and begin to course more caudolaterally, sweep around area 3b, and finally terminate in area 1/2. **b** Picture of the axons following injection in PMv in an animal that recovered from an M1 lesion. Fibers are observed between PMv and the lateral border of M1, as in control animals. In addition, atypical fibers are present lateral to the lesion. These fibers appear to originate from the border of M1 and are directed toward the parietal cortex. A *dark gray square* on the cartoon shows the approximate location of picture. **c** *Upper* reconstruction shows the pattern of terminal labeling observed in parietal areas in a control case. *Lower* reconstruction shows pattern of terminal labeling in an experimental case after recovery from the M1 lesion. A *dark gray square* on the cartoon show the approximate location of the reconstructions. Outlines of area 3b representations of digits and palm are shown. *Arrowheads* indicate the hand–face septum. *Black downward arrows* point to the area 1/2 hand representation. *Dotted line* indicates the approximate location of the lateral fissure. Scale bar 1 mm. **d** Quantification of the distribution of labeled terminals in areas 3a, 3b, and 1/2 of control and experimental cases. *Asterisks* show statistically significant differences ($p < 0.05$). (Adapted from Dancause et al. 2005)

relatively small craniotomy and facilitates the study of both areas. In addition, the larger DFL representation in PMv could facilitate the detection of subtle changes of surface area during recovery using ICMS techniques. However, other premotor areas share many of the same physiological and anatomical features as PMv, making them equally well suited to take over lost function after a lesion in M1. Not surprisingly, human studies have reported changes in CMA, SMA, PMd, and PMv in association with functional recovery (Carey et al. 2005, 2006; Fridman et al. 2004; Loubinoux et al. 2003; Miyai et al. 2003; Seitz et al. 2005).

Middle cerebral artery occlusions typically result in damage in the lateral sensorimotor cortex and spare medial areas, which are in part supplied by the anterior cerebral artery. As a consequence, in many stroke patients, SMA and the CMAs may be the only premotor areas available to support recovery of lost function.

To investigate the physiological reorganization of SMA, large ischemic lesions extending across M1, PMd, and PMv were induced in squirrel monkeys (Eisner-Janowicz et al. 2008). Following lesion, monkeys switched their hand preference and had substantial decrease of function with the impaired arm and hand. Some behavioral recovery occurred in the first 3 weeks postinjury, but animals had important residual deficits that remained relatively constant throughout the next 10 weeks of recovery. Analyses of the physiological organization of SMA showed an initial trend for reduction at week 3 that was followed by an expansion of the DFL representation, which became substantially larger than preinfarct values by week 13. At that time point, much similar to what we described above for PMv following lesions limited to M1, a linear relationship between the size of the SMA DFL area and the volume of the lesion was found. Although there was not a tight temporal match between the increase of the SMA DFL area and the recovery, motor performance at week 13 was positively correlated to the size of the DFL area in SMA. The reason for the temporal mismatch between the changes in cortical organization and motor performance is not clear, but a similar phenomenon is also known to occur with learning in normal animals (Kleim et al. 2004).

The functional implication of SMA in recovery of hand function was also supported by a study of the changes of single-neuron activity following recovery from an M1 lesion (Aizawa et al. 1991). In this study, a monkey was trained to perform visually instructed finger flexion with the right or left digits. Neurons in SMA typically show activity prior to movement production in this task (Tanji et al. 1988). However, in the overly trained monkey, very few neurons showed this premovement activity, suggesting that the automatization of the task performance was accompanied by a reduction of the activity of SMA neurons prior to movement. An electrolytic lesion induced in the M1 DFL resulted in marked deficits of behavioral performance and abnormal patterns of muscle activation, which progressively recovered to prelesion levels by 21 days. Twenty-two days after the lesion, neural recording revealed marked differences in the population of SMA neurons. Similarly to animals that were not overtrained on the task, many neurons in SMA were modulated prior to movement. It could be hypothesized that the relearning of the task after lesion was sustained by a reorganization of SMA neural activity and a return of its involvement in motor performance.

Reorganization of corticospinal projections of SMA DFL has been investigated in macaque monkeys following lesions that also included both M1 and the lateral premotor cortex (parts of PMd and PMv). After 6–12 months of recovery, an anterograde anatomical tracer was injected in SMA DFL. It was found that recovery was accompanied with an increase of projections from the ipsilesional SMA to the contralateral spinal cord laminae VII and IX (McNeal et al. 2010). Furthermore, these anatomical changes were correlated with functional recovery. In one animal, a secondary lesion in the ipsilesional SMA was performed and reinstated the motor deficits. These data support that areas remote from cortical injury can increase their corticospinal projections on the dennervated motor neurons in order to vicariously take over the lost function.

The new functional role of surviving premotor areas in recovery has also been supported by secondary inactivation experiments (Liu and Rouiller 1999). Two macaque monkeys that were trained to perform a precision-grip task sustained a permanent lesion of the sensorimotor cortex, including both M1 and part of the S1 hand representation. Monkeys progressively regained partial function of the hand in the following weeks and months. Nine months following the lesion a GABA-A agonist was injected to transiently inactivate different cortical areas of the ipsi- and contralesional hemisphere in different experimental sessions. Isolated inactivation of the contralesional M1, the ipsi- or contralesional PMd, and the ipsi- or contralesional PMv did not induce significant behavioral deficits of the recovered hand, but simultaneous inactivation of both the ipsilesional PMd and PMv reinstated the deficits caused by the lesion. These data support the hypothesis that both of these areas played an increased role in recovery from the lesion and the generation of movement.

Based on the data that we have reviewed, it appears that many premotor areas can support motor recovery after a lesion involving M1. It is likely that, the lack of evidence for the role of the CMAs in animal models to date is due to the technical challenges for the investigation of these areas, buried within the medial wall. Changes in all premotor areas, as well as in other areas involved in the motor network, conceivably occur in parallel to support recovery. The differential contribution of each premotor area to recovery remains unclear, and each premotor area might play a different role depending on the size and location of the lesion.

Concluding Statements

Stroke injuries pose a heavy burden on health care systems worldwide, with 15 million people suffering from stroke each year and 5 million of these patients living with residual disabilities (see Heart and Stroke foundation). With our aging population, these numbers are expected to increase. Now, there is extensive evidence that stroke survivors recover, at least in part, through adaptive plasticity of the surviving neural tissue. Functional recovery may thus be increased through a maximization of neural compensation, a concept that is being intensely investigated in the thriving field of neurorehabilitation research. Indeed, many novel approaches to facilitate plasticity and increase function are currently being designed and tested. The development of the field of neurorehabilitation and its novel therapies has capitalized on recent advances in our understandings of adaptive plasticity and neural mechanisms supporting recovery.

Lesions in animal models of stroke are generally much more restricted than those resulting from typical stroke in humans. In the present review, we have discussed many studies of plasticity and recovery following lesions targeting M1. As in humans every stroke is different in size and location, obviously these experimental lesions do not perfectly replicate the damage found in the entire population of stroke patients. However, these studies help to reveal more clearly the contribution of M1 lesions to the overall condition and highlight some of the mechanisms underlying recovery in the wide population of patients that have analogous cortical damage and deficits. The use of nonhuman primate models to investigate the vicarious role of diverse cortical specializations in recovery is especially useful. In particular, monkeys have a premotor organization that shares many similarities to humans (Crammond and Kalaska 1994; Kurata and Wise 1988; Luppino et al. 1993; Morecraft and Van Hoesen 1992; Murata et al. 1997; Nudo and Masterton 1989; Picard and Strick 1996; Scott et al. 1997) and thus provide an excellent model for investigating how these areas can contribute to recovery. Whereas the work in animal models reviewed here cannot immediately be translated to clinical practice, it has an impact on clinical research. These experiments contribute to our understanding of sensorimotor organization of the cortex as well as the capacity of this system to reorganize following injury. These data help to adjust hypotheses and treatment protocols currently being developed in clinical research. Ultimately, evidence-based protocols translate to advances in clinical practice and improvement in patients' quality of life after stroke.

Acknowledgments The author wishes to extend grateful thanks to Dr. Kelsey D. Dancause for grammar and insightful editing and Dr. Allan Smith for suggestions on scientific content. Numa Dancause is currently holding a Chercheur Boursier Junior 1 salary award from the Fonds de la Recherche en Santé du Québec (FRSQ) and a New Investigator salary award from the Canadian Institutes of Health Research (CIHR).

References

Aizawa H, Inase M, Mushiake H, Shima K, Tanji J (1991) Reorganization of activity in the supplementary motor area associated with motor learning and functional recovery. Exp Brain Res 84(3):668–671

Alexander GE, Crutcher MD (1990) Preparation for movement: neural representations of intended direction in three motor areas of the monkey. J Neurophysiol 64(1):133–150

Alstermark B, Kummel H, Pinter MJ, Tantisira B (1990) Integration in descending motor pathways controlling the forelimb in the cat. 17. Axonal projection and termination of C3-C4 propriospinal neurones in the C6-Th1 segments. Exp Brain Res 81(3):447–461

Asanuma H, Pavlides C (1997) Neurobiological basis of motor learning in mammals. Neuroreport 8(4):i–vi

Barbas H, Pandya DN (1987) Architecture and frontal cortical connections of the premotor cortex (area 6) in the rhesus monkey. J Comp Neurol 256(2):211–228
Bates JF, Goldman-Rakic PS (1993) Prefrontal connections of medial motor areas in the rhesus monkey. J Comp Neurol 336(2):211–228
Bennett KM, Lemon RN (1996) Corticomotoneuronal contribution to the fractionation of muscle activity during precision grip in the monkey. J Neurophysiol 75(5):1826–1842
Bortoff GA, Strick PL (1993) Corticospinal terminations in two new-world primates: further evidence that corticomotoneuronal connections provide part of the neural substrate for manual dexterity. J Neurosci 13(12):5105–5118
Boudrias M-H, Cheney PD (2006) Output properties of premotor ventral area (PMv) in rhesus macaques. Paper presented at the society for the neural control of movement, 16th annual meeting, Key Biscayne, FL
Brinkman C (1984) Supplementary motor area of the monkey's cerebral cortex: short- and long-term deficits after unilateral ablation and the effects of subsequent callosal section. J Neurosci 4(4):918–929
Buys EJ, Lemon RN, Mantel GW, Muir RB (1986) Selective facilitation of different hand muscles by single corticospinal neurones in the conscious monkey. J Physiol 381:529–549
Carey LM, Abbott DF, Egan GF, Bernhardt J, Donnan GA (2005) Motor impairment and recovery in the upper limb after stroke. Behavioral and neuroanatomical correlates. Stroke 36:625–629
Carey LM, Abbott DF, Egan GF, O'Keefe GJ, Jackson GD, Bernhardt J et al (2006) Evolution of brain activation with good and poor motor recovery after stroke. Neurorehabil Neural Repair 20(1):24–41
Carmichael ST, Archibeque I, Luke L, Nolan T, Momiy J, Li S (2005) Growth-associated gene expression after stroke: evidence for a growth-promoting region in peri-infarct cortex. Exp Neurol 193(2):291–311
Cerri G, Shimazu H, Maier MA, Lemon RN (2003) Facilitation from ventral premotor cortex of primary motor cortex outputs to macaque hand muscles. J Neurophysiol 90(2):832–842
Cheney PD, Fetz EE (1985) Comparable patterns of muscle facilitation evoked by individual corticomotoneuronal (CM) cells and by single intracortical microstimuli in primates: evidence for functional groups of CM cells. J Neurophysiol 53(3):786–804
Crammond DJ, Kalaska JF (1994) Modulation of preparatory neuronal activity in dorsal premotor cortex due to stimulus-response compatibility. J Neurophysiol 71(3):1281–1284
Dancause N, Barbay S, Frost SB, Plautz EJ, Chen D, Zoubina EV et al (2005) Extensive cortical rewiring after brain injury. J Neurosci 25(44):10167–10179
Dancause N, Barbay S, Frost SB, Plautz EJ, Popescu M, Dixon PM et al (2006a) Topographically divergent and convergent connectivity between premotor and primary motor cortex. Cereb Cortex 16(8):1057–1068
Dancause N, Barbay S, Frost SB, Plautz EJ, Stowe AM, Friel KM et al (2006b) Ipsilateral connections of the ventral premotor cortex in a new world primate. J Comp Neurol 495(4):374–390
Dancause N, Barbay S, Frost SB, Zoubina EV, Plautz EJ, Mahnken JD et al (2006c) Effects of small ischemic lesions in the primary motor cortex on neurophysiological organization in ventral premotor cortex. J Neurophysiol 96(6):3506–3511
Dancause N, Barbay S, Frost SB, Mahnken JD, Nudo RJ (2007) Interhemispheric connections of the ventral premotor cortex in a new world primate. J Comp Neurol 505(6):701–715
Dancause N, Duric V, Barbay S, Frost SB, Stylianou A, Nudo RJ (2008) An additional motor-related field in the lateral frontal cortex of squirrel monkeys. Cereb Cortex 18(12):2719–2728
Davare M, Lemon R, Olivier E (2008) Selective modulation of interactions between ventral premotor cortex and primary motor cortex during precision grasping in humans. J Physiol 586(Pt 11):2735–2742
Dum RP, Strick PL (1991) The origin of corticospinal projections from the premotor areas in the frontal lobe. J Neurosci 11(3):667–689
Dum RP, Strick PL (2002) Motor areas in the frontal lobe of the primate. Physiol Behav 77(4–5):677–682

Dum RP, Strick PL (2005) Frontal lobe inputs to the digit representations of the motor areas on the lateral surface of the hemisphere. J Neurosci 25(6):1375–1386
Ebner TJ, Hendrix CM, Pasalar S (2009) Past, present, and emerging principles in the neural encoding of movement. Adv Exp Med Biol 629:127–137
Eisner-Janowicz I, Barbay S, Hoover E, Stowe AM, Frost SB, Plautz EJ et al (2008) Early and late changes in the distal forelimb representation of the supplementary motor area after injury to frontal motor areas in the squirrel monkey. J Neurophysiol 100(3):1498–1512
Fang PC, Stepniewska I, Kaas JH (2005) Ipsilateral cortical connections of motor, premotor, frontal eye, and posterior parietal fields in a prosimian primate, Otolemur garnetti. J Comp Neurol 490(3):305–333
Fetz EE, Cheney PD (1980) Postspike facilitation of forelimb muscle activity by primate corticomotoneuronal cells. J Neurophysiol 44(4):751–772
Fetz EE, Cheney PD, German DC (1976) Corticomotoneuronal connections of precentral cells detected by postspike averages of EMG activity in behaving monkeys. Brain Res 114(3):505–510
Florence SL, Taub HB, Kaas JH (1998) Large-scale sprouting of cortical connections after peripheral injury in adult macaque monkeys. Science 282(5391):1117–1121
Fogassi L, Gallese V, Buccino G, Craighero L, Fadiga L, Rizzolatti G (2001) Cortical mechanism for the visual guidance of hand grasping movements in the monkey: a reversible inactivation study. Brain 124(Pt 3):571–586
Fridman EA, Hanakawa T, Chung M, Hummel F, Leiguarda RC, Cohen LG (2004) Reorganization of the human ipsilesional premotor cortex after stroke. Brain 127(Pt 4):747–758
Friel KM, Barbay S, Frost SB, Plautz EJ, Hutchinson DM, Stowe AM et al (2005) Dissociation of sensorimotor deficits after rostral versus caudal lesions in the primary motor cortex hand representation. J Neurophysiol 94(2):1312–1324
Friel KM, Barbay S, Frost SB, Plautz EJ, Stowe AM, Dancause N et al (2007) Effects of a rostral motor cortex lesion on primary motor cortex hand representation topography in primates. Neurorehabil Neural Repair 21(1):51–61
Frost SB, Barbay S, Friel KM, Plautz EJ, Nudo RJ (2003) Reorganization of remote cortical regions after ischemic brain injury: a potential substrate for stroke recovery. J Neurophysiol 89(6): 3205–3214
Fulton J (1935) A note on the definition of the "motor" and "premotor" areas. Brain 58:311–316
Gentilucci M, Fogassi L, Luppino G, Matelli M, Camarda R, Rizzolatti G (1988) Functional organization of inferior area 6 in the macaque monkey. I. Somatotopy and the control of proximal movements. Exp Brain Res 71(3):475–490
Georgopoulos AP, Stefanis CN (2007) Local shaping of function in the motor cortex: motor contrast, directional tuning. Brain Res Rev 55(2):383–389
Gerbella M, Belmalih A, Borra E, Rozzi S, Luppino G (2011). Cortical connections of the anterior (F5a) subdivision of the macaque ventral premotor area F5. Brain Struct Funct 216(1):43–65
Geyer S, Matelli M, Luppino G, Zilles K (2000) Functional neuroanatomy of the primate isocortical motor system. Anat Embryol (Berl) 202(6):443–474
Ghosh S, Gattera R (1995) A comparison of the ipsilateral cortical projections to the dorsal and ventral subdivisions of the macaque premotor cortex. Somatosens Mot Res 12(3–4):359–378
Glees P, Cole J (1950) Recovery of skilled motor functions after small repeated lesions of motor cortex in macaque. J Neurophysiol 13:137–148
Griffin DM, Hudson HM, Belhaj-Saif A, Cheney PD (2009) Stability of output effects from motor cortex to forelimb muscles in primates. J Neurosci 29(6):1915–1927
He SQ, Dum RP, Strick PL (1993) Topographic organization of corticospinal projections from the frontal lobe: motor areas on the lateral surface of the hemisphere. J Neurosci 13(3):952–980
He SQ, Dum RP, Strick PL (1995) Topographic organization of corticospinal projections from the frontal lobe: motor areas on the medial surface of the hemisphere. J Neurosci 15(5 Pt 1): 3284–3306
Hepp-Reymond MC, Husler EJ, Maier MA, Ql HX (1994) Force-related neuronal activity in two regions of the primate ventral premotor cortex. Can J Physiol Pharmacol 72(5):571–579

Huntley GW, Jones EG (1991) Relationship of intrinsic connections to forelimb movement representations in monkey motor cortex: a correlative anatomic and physiological study. J Neurophysiol 66(2):390–413
Kaas JH (1993) The functional organization of somatosensory cortex in primates. Anat Anz 175(6):509–518
Kalaska JF (2009) From intention to action: motor cortex and the control of reaching movements. Adv Exp Med Biol 629:139–178
Kalaska JF, Sergio LE, Cisek P (1998) Cortical control of whole-arm motor tasks. Novartis Found Symp 218:176–190; discussion 190–201
Kermadi I, Liu Y, Tempini A, Rouiller EM (1997) Effects of reversible inactivation of the supplementary motor area (SMA) on unimanual grasp and bimanual pull and grasp performance in monkeys. Somatosens Mot Res 14(4):268–280
Kleim JA, Hogg TM, VandenBerg PM, Cooper NR, Bruneau R, Remple M (2004) Cortical synaptogenesis and motor map reorganization occur during late, but not early, phase of motor skill learning. J Neurosci 24(3):628–633
Kubota K (1996) Motor cortical muscimol injection disrupts forelimb movement in freely moving monkeys. Neuroreport 7(14):2379–2384
Kurata K, Hoffman DS (1994) Differential effects of muscimol microinjection into dorsal and ventral aspects of the premotor cortex of monkeys. J Neurophysiol 71(3):1151–1164
Kurata K, Wise SP (1988) Premotor and supplementary motor cortex in rhesus monkeys: neuronal activity during externally- and internally-instructed motor tasks. Exp Brain Res 72(2):237–248
Lai SM, Studenski S, Duncan PW, Perera S (2002) Persisting consequences of stroke measured by the Stroke Impact Scale. Stroke 33(7):1840–1844
Lashley KS (1929) Brain mechanisms and intelligence: a quantitative study of injuries to the brain. Chicago Press, Chicago
Lashley KS (1930) Basic neural mechanisms in behavior. Psychol Rev 37:1–24
Lemon RN, Muir RB, Mantel GW (1987) The effects upon the activity of hand and forearm muscles of intracortical stimulation in the vicinity of corticomotor neurones in the conscious monkey. Exp Brain Res 66(3):621–637
Liu Y, Rouiller EM (1999) Mechanisms of recovery of dexterity following unilateral lesion of the sensorimotor cortex in adult monkeys. Exp Brain Res 128(1–2):149–159
Loubinoux I, Carel C, Pariente J, Dechaumont S, Albucher JF, Marque P et al (2003) Correlation between cerebral reorganization and motor recovery after subcortical infarcts. Neuroimage 20(4):2166–2180
Lu MT, Preston JB, Strick PL (1994) Interconnections between the prefrontal cortex and the premotor areas in the frontal lobe. J Comp Neurol 341(3):375–392
Luppino G, Matelli M, Camarda RM, Gallese V, Rizzolatti G (1991) Multiple representations of body movements in mesial area 6 and the adjacent cingulate cortex: an intracortical microstimulation study in the macaque monkey. J Comp Neurol 311(4):463–482
Luppino G, Matelli M, Camarda R, Rizzolatti G (1993) Corticocortical connections of area F3 (SMA-proper) and area F6 (pre-SMA) in the macaque monkey. J Comp Neurol 338(1):114–140
Maier MA, Olivier E, Baker SN, Kirkwood PA, Morris T, Lemon RN (1997) Direct and indirect corticospinal control of arm and hand motoneurons in the squirrel monkey (Saimiri sciureus). J Neurophysiol 78(2):721–733
Maier MA, Illert M, Kirkwood PA, Nielsen J, Lemon RN (1998) Does a C3-C4 propriospinal system transmit corticospinal excitation in the primate? An investigation in the macaque monkey. J Physiol 511(Pt 1):191–212
Matelli M, Luppino G, Rizzolatti G (1985) Patterns of cytochrome oxidase activity in the frontal agranular cortex of the macaque monkey. Behav Brain Res 18(2):125–136
Matelli M, Camarda R, Glickstein M, Rizzolatti G (1986) Afferent and efferent projections of the inferior area 6 in the macaque monkey. J Comp Neurol 251(3):281–298
Matelli M, Luppino G, Fogassi L, Rizzolatti G (1989) Thalamic input to inferior area 6 and area 4 in the macaque monkey. J Comp Neurol 280(3):468–488

Matelli M, Luppino G, Rizzolatti G (1991) Architecture of superior and mesial area 6 and the adjacent cingulate cortex in the macaque monkey. J Comp Neurol 311(4):445–462
Matsuzaka Y, Tanji J (1996) Changing directions of forthcoming arm movements: neuronal activity in the presupplementary and supplementary motor area of monkey cerebral cortex. J Neurophysiol 76(4):2327–2342
Matsuzaka Y, Aizawa H, Tanji J (1992) A motor area rostral to the supplementary motor area (presupplementary motor area) in the monkey: neuronal activity during a learned motor task. J Neurophysiol 68(3):653–662
McKiernan BJ, Marcario JK, Karrer JH, Cheney PD (1998) Corticomotoneuronal postspike effects in shoulder, elbow, wrist, digit, and intrinsic hand muscles during a reach and prehension task. J Neurophysiol 80(4):1961–1980
McNeal DW, Darling WG, Ge J, Stilwell-Morecraft KS, Solon KM, Hynes SM et al (2010) Selective long-term reorganization of the corticospinal projection from the supplementary motor cortex following recovery from lateral motor cortex injury. J Comp Neurol 518(5):586–621
Miyai I, Yagura H, Hatakenaka M, Oda I, Konishi I, Kubota K (2003) Longitudinal optical imaging study for locomotor recovery after stroke. Stroke 34(12):2866–2870
Morecraft RJ, Van Hoesen GW (1992) Cingulate input to the primary and supplementary motor cortices in the rhesus monkey: evidence for somatotopy in areas 24c and 23c. J Comp Neurol 322(4):471–489
Munk H (1881) Uber die Funktionen der Grosshirnrinde. In: Hirshwald A (ed) Gesammelte Mitteilungen aus den Jahren. Hirshwald, Berlin, pp 1877–1880
Murata A, Fadiga L, Fogassi L, Gallese V, Raos V, Rizzolatti G (1997) Object representation in the ventral premotor cortex (area F5) of the monkey. J Neurophysiol 78(4):2226–2230
Nudo RJ, Masterton RB (1989) Descending pathways to the spinal cord: II. Quantitative study of the tectospinal tract in 23 mammals. J Comp Neurol 286(1):96–119
Nudo RJ, Milliken GW (1996a) Reorganization of movement representations in primary motor cortex following focal ischemic infarcts in adult squirrel monkeys. J Neurophysiol 75(5): 2144–2149
Nudo RJ, Milliken GW, Jenkins WM, Merzenich MM (1996b) Use-dependent alterations of movement representations in primary motor cortex of adult squirrel monkeys. J Neurosci 16(2):785–807
Nudo RJ, Wise BM, SiFuentes F, Milliken GW (1996c) Neural substrates for the effects of rehabilitative training on motor recovery after ischemic infarct. Science 272(5269):1791–1794
Ogden R, Franz SI (1917) On cerebral motor control: the recovery of function from experimentally produced hemiplegia. Psychbiol 1:33–50
Park MC, Belhaj-Saif A, Gordon M, Cheney PD (2001) Consistent features in the forelimb representation of primary motor cortex in rhesus macaques. J Neurosci 21(8):2784–2792
Penfield W, Boldrey E (1937). Somatic motor and sensory representation in the cerebreal cortex of man as studied by electrical stimulation. Brain 60:389–443
Phillips CG, Porter R (1977) Corticospinal neurones. Their role in movement. Monogr Physiol Soc 34:v–xii, 1–450
Picard N, Strick PL (1996) Motor areas of the medial wall: a review of their location and functional activation. Cereb Cortex 6(3):342–353
Picard N, Strick PL (2001) Imaging the premotor areas. Curr Opin Neurobiol 11(6):663–672
Pierrot-Deseilligny E (1996) Transmission of the cortical command for human voluntary movement through cervical propriospinal premotoneurons. Prog Neurobiol 48(4–5):489–517
Pineiro R, Pendlebury ST, Smith S, Flitney D, Blamire AM, Styles P et al (2000) Relating MRI changes to motor deficit after ischemic stroke by segmentation of functional motor pathways. Stroke 31(3):672–679
Plautz EJ, Milliken GW, Nudo RJ (2000) Effects of repetitive motor training on movement representations in adult squirrel monkeys: role of use versus learning. Neurobiol Learn Mem 74(1):27–55

Porter R (1985) The corticomotoneuronal component of the pyramidal tract: corticomotoneuronal connections and functions in primates. Brain Res 357(1):1–26
Prabhu G, Shimazu H, Cerri G, Brochier T, Spinks RL, Maier MA et al (2009) Modulation of primary motor cortex outputs from ventral premotor cortex during visually guided grasp in the macaque monkey. J Physiol 587(Pt 5):1057–1069
Preuss TM, Stepniewska I, Kaas JH (1996) Movement representation in the dorsal and ventral premotor areas of owl monkeys: a microstimulation study. J Comp Neurol 371(4):649–676
Ralston DD, Ralston HJ 3rd (1985) The terminations of corticospinal tract axons in the macaque monkey. J Comp Neurol 242(3):325–337
Rathelot JA, Strick PL (2006) Muscle representation in the macaque motor cortex: an anatomical perspective. Proc Natl Acad Sci U S A 103(21):8257–8262
Rathelot JA, Strick PL (2009) Subdivisions of primary motor cortex based on cortico-motoneuronal cells. Proc Natl Acad Sci U S A 106(3):918–923
Recanzone GH, Merzenich MM, Jenkins WM (1992a) Frequency discrimination training engaging a restricted skin surface results in an emergence of a cutaneous response zone in cortical area 3a. J Neurophysiol 67(5):1057–1070
Recanzone GH, Jenkins WM, Hradek GT, Merzenich MM (1992b) Progressive improvement in discriminative abilities in adult owl monkeys performing a tactile frequency discrimination task. J Neurophysiol 67(5):1015–1030
Recanzone GH, Merzenich MM, Jenkins WM, Grajski KA, Dinse HR (1992c) Topographic reorganization of the hand representation in cortical area 3b owl monkeys trained in a frequency-discrimination task. J Neurophysiol 67(5):1031–1056
Rizzolatti G, Matelli M, Pavesi G (1983) Deficits in attention and movement following the removal of postarcuate (area 6) and prearcuate (area 8) cortex in macaque monkeys. Brain 106(Pt 3): 655–673
Rizzolatti G, Camarda R, Fogassi L, Gentilucci M, Luppino G, Matelli M (1988) Functional organization of inferior area 6 in the macaque monkey. II. Area F5 and the control of distal movements. Exp Brain Res 71(3):491–507
Rizzolatti G, Gentilucci M, Camarda RM, Gallese V, Luppino G, Matelli M et al (1990) Neurons related to reaching-grasping arm movements in the rostral part of area 6 (area 6a beta). Exp Brain Res 82(2):337–350
Rouiller EM, Babalian A, Kazennikov O, Moret V, Yu XH, Wiesendanger M (1994) Transcallosal connections of the distal forelimb representations of the primary and supplementary motor cortical areas in macaque monkeys. Exp Brain Res 102(2):227–243
Sakai ST, Stepniewska I, Qi HX, Kaas JH (2000) Pallidal and cerebellar afferents to pre-supplementary motor area thalamocortical neurons in the owl monkey: a multiple labeling study. J Comp Neurol 417(2):164–180
Schieber MH (2000) Inactivation of the ventral premotor cortex biases the laterality of motoric choices. Exp Brain Res 130(4):497–507
Schieber MH (2001) Constraints on somatotopic organization in the primary motor cortex. J Neurophysiol 86(5):2125–2143
Schieber MH, Hibbard LS (1993) How somatotopic is the motor cortex hand area? Science 261(5120):489–492
Schieber MH, Poliakov AV (1998) Partial inactivation of the primary motor cortex hand area: effects on individuated finger movements. J Neurosci 18(21):9038–9054
Schmidlin E, Brochier T, Maier MA, Kirkwood PA, Lemon RN (2008) Pronounced reduction of digit motor responses evoked from macaque ventral premotor cortex after reversible inactivation of the primary motor cortex hand area. J Neurosci 28(22):5772–5783
Scott SH, Sergio LE, Kalaska JF (1997) Reaching movements with similar hand paths but different arm orientations. II. Activity of individual cells in dorsal premotor cortex and parietal area 5. J Neurophysiol 78(5):2413–2426
Seitz RJ, Kleiser R, Butefisch CM (2005) Reorganization of cerebral circuits in human brain lesion. Acta Neurochir Suppl 93:65–70

Shima K, Mushiake H, Saito N, Tanji J (1996) Role for cells in the presupplementary motor area in updating motor plans. Proc Natl Acad Sci U S A 93(16):8694–8698

Shimazu H, Maier MA, Cerri G, Kirkwood PA, Lemon RN (2004) Macaque ventral premotor cortex exerts powerful facilitation of motor cortex outputs to upper limb motoneurons. J Neurosci 24(5):1200–1211

Shinoda Y, Zarzecki P, Asanuma H (1979) Spinal branching of pyramidal tract neurons in the monkey. Exp Brain Res 34(1):59–72

Shinoda Y, Futami T, Kano M (1985) Synaptic organization of the cerebello-thalamo-cerebral pathway in the cat. II. Input-output organization of single thalamocortical neurons in the ventrolateral thalamus. Neurosci Res 2(3):157–180

Shinoda Y, Yamaguchi T, Futami T (1986) Multiple axon collaterals of single corticospinal axons in the cat spinal cord. J Neurophysiol 55(3):425–448

Stepniewska I, Preuss TM, Kaas JH (1993) Architectonics, somatotopic organization, and ipsilateral cortical connections of the primary motor area (M1) of owl monkeys. J Comp Neurol 330(2):238–271

Strick PL, Preston JB (1982) Two representations of the hand in area 4 of a primate. II. Somatosensory input organization. J Neurophysiol 48(1):150–159

Takada M, Tokuno H, Hamada I, Inase M, Ito Y, Imanishi M et al (2001) Organization of inputs from cingulate motor areas to basal ganglia in macaque monkey. Eur J Neurosci 14(10):1633–1650

Tanji J, Okano K, Sato KC (1988) Neuronal activity in cortical motor areas related to ipsilateral, contralateral, and bilateral digit movements of the monkey. J Neurophysiol 60(1):325–343

Vogt C, Vogt O (1919) Allgemeinere Ergebnisse unserer Hirnforschung. J Psychol Neurol (Leipz.) 25:279–461

Von Bonin G, Bailey P (1947) The neocortex of *Macaca Mulatta*, vol 136. University of Illinois Press, Urbana

Wiesendanger R, Wiesendanger M (1985) The thalamic connections with medial area 6 (supplementary motor cortex) in the monkey (macaca fascicularis). Exp Brain Res 59(1):91–104

Woolsey CN, Settlage PH, Meyer DR, Spencer W, Hamuy TP, Travis AM (1952) Patterns of localization in precentral and "supplementary" motor areas and their relation to the concept of a premotor area. Res Publ Assoc Res Nerv Ment Dis 30:238–264

Zhu LL, Lindenberg R, Alexander MP, Schlaug G (2010) Lesion load of the corticospinal tract predicts motor impairment in chronic stroke. Stroke 41(5):910–915

Chapter 5
The Mirror System in Monkeys and Humans and its Possible Motor-Based Functions

Leonardo Fogassi and Luciano Simone

Introduction

Classically, the caudal agranular sector of the frontal lobe of primates has been considered to be involved in motor functions, whereas the rostral one was assigned mainly a role in "cognitive" processes. This view was the result of major electrophysiological studies on the cerebral cortex (Woolsey et al. 1952; Evarts 1968) supporting the idea that the functional role of the caudal frontal lobe was that of producing body-part movements. However, more recent behavioral and psychophysical studies (Jeannerod 1988; Rosenbaum et al. 2007) and neurophysiological experiments (see Fogassi and Ferrari 2011) showed that the motor system has different levels of organization, each of which participates in collaboration with the others, to the achievement of the intended goal of an action. The first (highest) level is the 'action' level. An action is composed of a sequence of fluently linked motor acts leading to the accomplishment of a final behavioral goal. Motor acts, that constitute the second (middle) level, are movements aimed at an intermediate goal (e.g., grasping an object or bringing a piece of food to the mouth); each motor act can be involved in the formation of different actions expressing different final goals. The act of 'movement' constitutes the third level. Movements can be defined as displacements of single joints. Each motor act is composed of two or more synergic movements, and the organization of motor act sequences produce organized goal-directed actions in space. Within the cortex, different sectors of the frontal lobe contribute to the encoding of most of these levels. Available data indicate that the action level can depend on the activity of prefrontal cortex and can partly be expressed in higher order motor areas. The neurons contained in these latter areas (together forming area 6 of Brodmann, BA6) mostly code motor acts, whereas movements are mainly coded by primary

L. Fogassi (✉) · L. Simone
Dipartimento di Neuroscienze e Istituto Italiano di Tecnologia (RTM),
Università di Parma, Parma, Italy
e-mail: leonardo.fogassi@unipr.it

L. Fogassi
Dipartimento di Psicologia, Università di Parma, Parma, Italy

M. J. Richardson et al. (eds.), *Progress in Motor Control*,
Advances in Experimental Medicine and Biology 782,
DOI 10.1007/978-1-4614-5465-6_5,

motor cortex (corresponding to area 4 of Brodmann, BA4). The actual movement execution depends on the activation of spinal motor neurons that determine muscles contraction.

The abstract level of goal coding present in cortical motor areas covers functions traditionally attributed to the classical associative areas. This view is further supported by the results obtained in neuroanatomical studies that highlight the existence of parieto-premotor circuits in which specific areas of the posterior parietal cortex are reciprocally linked with specific areas of the motor cortex (Rizzolatti and Luppino 2001). The functional correlate of these circuits is the transformation of sensory stimuli into appropriate motor acts (Rizzolatti et al. 1997). The existence of these connections challenged the classical view of separate sensory and motor functions, indicating that perception and action share the same neuronal substrates. A very intriguing example of this action–perception coupling is the discovery of a system linking action observation with action production, namely the mirror neuron system.

We will first describe the main features of mirror neurons in monkeys, followed by some of the evidence on the mirror system in humans. Finally, we will show how this system can be involved in social cognitive functions.

Mirror Neurons in Monkeys

Mirror neurons were first discovered in a sector of the macaque ventral premotor cortex (area F5) (Di Pellegrino et al. 1992; Gallese et al. 1996; Rizzolatti et al. 1996a) and subsequently were also found in the inferior parietal lobule (area PFG) (Gallese et al. 2002; Fogassi et al. 2005; Rozzi et al. 2008). They represent a class of visuomotor neurons that discharge both when the monkey *performs* a given motor act (e.g., grasping, manipulating, tearing an object) and when it *observes* the same, or a similar, motor act done by another individual. In order to be activated, these neurons require an interaction between a biological effector (hand or mouth) and an object. In fact the presentation of the target on which the motor act is performed, a person mimicking a motor act, or an individual making intransitive (nonobject-directed) gestures are ineffective in eliciting their response. The observed hand motor acts that are more effective in producing mirror neurons response are grasping, manipulating, holding, and other hands interactions. Among them, grasping is the most represented motor act. Part of mirror neurons respond to the observation of only one motor act, whereas others respond during the observation of two or three motor acts. The mirror neuron response is largely invariant with respect to many factors that, in principle, could modulate their discharge, such as the sector of space in which the motor act is performed and the size and type of object target of the motor act, etc. However, a small percentage of mirror neurons can be differently modulated by the observed right or left hand performing the motor act, regardless of whether the observed act is executed in front of the monkey, on its left side, or on its right side. Note that a similar hand preference activity, not necessarily matching with the visual response preference, can be exhibited also during monkey's grasping execution. Another small group of mirror neurons can also be modulated by the direction of the observed motor act (i.e., the experimenter performing the action from the left to the right or in the opposite direction).

A category of mirror neurons respond to the observation of mouth motor acts (mouth mirror neurons). Mouth mirror neurons have been mainly described in the lateral part of area F5 and in area PF/PFG. Most of them are responsive during the observation and execution of ingestive motor acts such as biting, sucking, licking, etc. (Ferrari et al. 2003). A small portion of mouth mirror neurons also respond to the observation of mouth communicative gestures, such as lips-smacking or tongue protrusion. This subset of mouth mirror neurons could represent an old system of communication, likely evolved from neurons expressing different types of goal (ingestion).

When mirror neurons are categorized on the basis of the *'congruence'* between the executed and the observed motor act effective in activating them, it is possible to define two different sets of neurons that is "strictly congruent" and "broadly congruent" neurons. The strictly congruent neurons (about one third of mirror neurons) are neurons in which the executed and observed motor acts match both in terms of the goal (e.g., grasping) and the means to achieve them (e.g., whole hand grip). The other two thirds (broadly congruent) showed a congruence in terms of the goal, but a lower specificity for the observed act or the type of grip during observation as compared with execution (Gallese et al. 1996).

The congruence between the observed and the executed motor act represents the most important property of mirror neurons. It has been proposed that the "direct matching" of the visual description of another's motor act with one's own motor repertoire allows the observer to understand *what* another individual is doing. In fact, the observation of a motor act determines an automatic retrieval of the motor representation already available within the 'motor vocabulary' of the observer. An issue raised by the discovery of mirror neurons is why the activation of the motor representation obtained during motor act observation does not normally generate an overt motor output. This phenomenon is of course necessary for inhibiting immediate acts reproduction during the observation of conspecifics acting. However, it is not known what is the mechanism for this inhibition. Interestingly enough, Kraskov et al. (2009) recently demonstrated that half of F5 mirror neurons projecting to the spinal cord through the pyramidal tract were strongly excited during grasping execution, but their discharge was suppressed during grasping observation. The authors proposed that this inhibition could serve the purpose of reducing motor output response of mirror neurons during action observation.

The capability of mirror neurons to understand the purpose of motor events has been tested in different neurophysiological investigations. One study (Umiltà et al. 2001) was demonstrated that mirror neurons can discharge not only when the monkey can see a fully visible goal directed motor act but also when it sees only part of it—being the crucial part of the motor act (hand–target interaction)—hidden behind a screen. In a control experiment, it has also been shown that the neurons ceased discharging when the monkey knew that there was no object behind the screen. This result demonstrate that mirror neurons can retrieve the motor representation corresponding to the observed motor act, provided that prior contextual information and part of the observed act are available (object presence). This enables the observer to understand others' actions even if their full visual description is missing. In another study (Kohler et al. 2002), the authors showed that many mirror neurons that responded to the observation of motor acts that typically produce sound (for example, peanut breaking), also responded to the sound alone ("audio-visual" mirror neurons). The properties of this class of mirror neurons, represent the capacity of understanding the meaning of a motor act through different sensory modalities; it is noteworthy that this feature is also typical of language.

Although the early studies carried out on mirror neurons were focused on their motor and visual responses to motor acts performed with biological effectors (hand, mouth), more recently, two studies showed that mirror neurons can also be sensitive to motor acts done with artificial effectors (e.g., tools). In the first study, Umiltà et al. (2008) described single-neuron activity recorded in ventral premotor cortex from monkeys trained to grasp objects by means of two different types of pliers, direct and reverse pliers. The important feature was that the two types of pliers required an opposite pattern of finger movements (opening-closing or viceversa) in order to grasp a piece of food. Very interestingly, when tested while the monkey grasped food with the hand or with the pliers, motor neurons in area F5 discharged during the attainment of the goal (grasping the object) independently of the used effectors (either the hand or the pliers) and the pattern of finger movements used to achieve it (closing or opening of the fingers). Thus, the pliers, after learning, were coded in the motor system as if they were a prolongation of the hand. In fact, the same discharge was observed not only when the mechanics of pliers was congruent with that of the hand (normal pliers), but also when the mechanics was opposite (reverse pliers). The same type of response was also found during the observation of motor acts performed with the hand and both types of pliers. Interestingly, the observation of a motor act performed with another tool that the monkey was not trained to use elicited a lower discharge (Rochat et al. 2010). Altogether, these findings demonstrate that after motor learning mirror neurons can generalize their visual response, likely relying on the training-induced expansion of the motor repertoire.

It has been recently demonstrated (Ferrari et al. 2005) that the visual response of mirror neurons could not be strictly related to the monkeys' motor repertoire. These authors reported, a new type of visuomotor neurons in ventral premotor cortex (tool-responding mirror neurons) discharging during the observation of motor acts made with tools (a stick or pliers) after several months of visual exposure to an experimenter using these tools. When tested for the use of the observed tool, monkeys did not show any capacity to use it. The authors suggested that the long-lasting visual experience could create the establishment of a visual association between the tool and the experimenter's hand. In this way visual inputs related to stimuli that share similar motion and are directed to the same target, would be coded from F5 mirror neurons that already have a specific motor response (usually grasping with the hand and the mouth), but are still uncommitted on the visual side. From an ethological point of view the function of tool-responding mirror neurons could be that of extending the comprehension of the purpose of an action to other effectors, even when that action is not included among the motor representations of the observer but share with them similar motor goals (taking possession of an object).

Intrestingly, a recent functional magnetic resonance imaging (fMRI) study showed that in naïve and trained monkeys, the observation of grasping acts performed with simple tools activates a parieto-frontal circuit that is also active during the observation and execution of hand grasping movements (Peeters et al. 2009, see also below).

The early experiments carried out on mirror neurons were mainly focused on the matching between the visual and the motor response. However, more recent studies have shown that mirror neuron activity can be modulated by different features of the observed motor acts.

A recent study (Caggiano et al. 2009) showed that in addition to coding the goal of a motor act, the visual discharge of mirror neurons can be modulated by the space sector where the observed agent performs the motor act. This study was aimed to verify whether the discharge of mirror neurons can be modulated by the distance at which the observed act is executed. The results show that F5 mirror neurons can discharge differently depending on whether the observed motor act was performed by the other agent within the monkey's reaching space (peripersonal space) or outside it (extrapersonal space). In particular, 50 % of the studied mirror neurons discharged differently in the two conditions. Some mirror neurons fired stronger when the experimenter grasped a piece of food within the monkey peripersonal space with respect to the extrapersonal one. Other neurons behaved in the opposite way, coding others' action when performed in the extrapersonal space. Moreover, a subset of the space-sensitive mirror neurons could dynamically change their response, when a barrier was introduced between the monkey and the observed motor act. In this condition, extrapersonal neurons started discharging strongly also within the peripersonal space, while peripersonal neurons ceased discharging. This result suggests that space-sensitive mirror activity depends on the monkey's operational space. It has been suggested that the space-related differential response of mirror neurons represents the link between the comprehension of others' actions and the possibility to socially interact with the agent performing the action. For example, the discharge of mirror neurons preferring the peripersonal space could drive a competitive or (in the case of humans) cooperative reactions.

A recent investigation (Caggiano et al. 2011) highlights the capacity of mirror neurons to visually code goal-directed motor acts in a view-dependent way. Classically, the visual response of mirror neurons has been tested by using a naturalistic approach, with the experimenter executing goal-directed motor acts in front of the monkey. In this study, the activity of F5 mirror neurons were evaluated by presenting to the monkey movies showing grasping motor acts seen from different perspectives (frontal, lateral, and egocentric).

The results of this experiment showed that mirror neurons can also be activated when the observed motor act is presented in a movie, even though the response is generally weaker than in the naturalistic condition. Furthermore, using movies, it has been demonstrated that the response of most mirror neurons was modulated by the visual perspective from which a motor act was seen by an observer (Caggiano et al. 2011). In fact, only a small part of the studied neurons appeared to be view-independent, while the remaining appeared to encode, in equal percentage, the three different perspectives used in this study. These results suggest that, beyond encoding the goal of motor acts, mirror neurons can also contribute to provide the details on specific aspects of the observed act. It has been proposed that this latter function could be performed through the feedback connections linking the mirror neurons frontal and parietal areas with the temporal areas encoding the different pictorial views of a motor act (see also the section on the parieto-frontal mirror circuit).

Recently, several studies described neurons with properties similar to those of mirror neurons. Cisek and collaborators (Cisek and Kalaska 2005) have described a particular class of visuomotor neurons recorded in the dorsal premotor cortex (PMd) that became active both when the trained monkey was using a cursor to reach a target on a screen (motor response) and when the monkey observed the cursor, moved by the experimenter, reaching the same target (visual response). Note that in the visual condition the monkey sees the cursor, but not the experimenter's hand. Thus, different from classical mirror neurons, this class of neurons does not require the visual information about the effector used to achieve the target. Since they seem to code the goal of the observed cursor movement, it has been proposed that their discharge could be a rehearsal of the related motor act.

Another neurophysiological study (Tkach et al. 2007) described the response properties of neurons of primary motor cortex and dorsal premotor cortex (PMd) that became active when the monkey moved a cursor to reach a target on a screen and when it observed the same replayed movements. Interestingly, the visual response was present only when both the cursor and the target were visible. The presence of the replayed cursor movement without the target did not elicit the same effective neuronal response.

A further study (Shepherd et al. 2009) showed other mirror-like responses in the lateral intraparietal area (LIP). This area is known to play a crucial role in coding intended eye movements. These neurons activated both when the recorded monkey looked in the preferred direction of the LIP neuron and when the recorded monkey observed a picture of another monkey looking in the same direction. As proposed by the authors, this finding suggests a possible role of this mirror-like neurons in automatic social responses such as the gaze-following behavior contributing to the sharing of attention, considered as a fundamental step in social cognition (Baron-Cohen 1994).

The Parieto-Frontal Mirror Neuron Circuit

One of the main issues about mirror neuron properties still under dispute concerns the source of visual information to the mirror system. It has been reported that neurons recorded in the anterior region of the superior temporal sulcus (STSa) discharge during the observation of biological movements such as walking, head rotation, forelimb movements (Perrett et al. 1989; Barraclough et al. 2006). A small percentage of these neurons also discharged during the observation of hand goal-directed movements. STSa region cannot be strictly considered as a part of the mirror neuron system because its neurons do not exhibit *motor-related* discharge, but it is likely the source of visual information for this system. However, this STS region is not directly linked to premotor cortex, thus an intermediate node is required. As a matter of fact, mirror neurons having properties similar to that described in F5 have been recently described in the inferior parietal lobule (IPL) (Fogassi et al. 2005; Rozzi et al. 2008), more precisely in the cytoarchitectonic area PFG (Pandya and Seltzer 1982; Gregoriou et al. 2006). Thus, inferior parietal cortex can be considered as the first node of the mirror neuron system receiving visual information on biological stimuli.

Although the main findings of the presence of the mirror neuron mechanism has been provided by electrophysiological studies performed on specific cortical areas of the monkey, a complete picture of the areas involved in the action observation circuit can be better obtained from neuroimaging studies. Recently, two fMRI studies in the monkey provided evidence for the involvement of the parieto-frontal circuit in action observation. Both studies were performed in awake monkeys trained to fixate, while movies showing actions were presented to them. The first fMRI investigation (Nelissen et al. 2005) provided a functional parcellation of F5. In fact, the authors identified three different sectors of this area that were differently activated depending on whether the observed video represented a person grasping an object (acting person), or only a hand grasping objects (hand action). The results show that the convexity of F5 (F5c), where mirror neurons have been recorded, was more activated during the observation of the full view of a person executing a grasping act. The other two subdivisions, F5a and F5p, corresponding to the anterior and posterior sector of the arcuate sulcus, respectively, were more activated during the observation of a grasping hand, thus appearing to code actions in a less context-dependent way.

The second fMRI experiment (Nelissen et al. 2011), conducted using the same visual stimuli as the previous one, gave two main results. The first was that, in analogy with the findings of the previous study, the area PFG was more active during observation of the whole person grasping an object, while area AIP preferred the view of the hand grasping. The second is that during the observation of grasping motor acts different areas of STS, both in its upper and lower banks, became active. Hodological studies showed that those areas are anatomically connected with different areas of IPL (Rozzi et al. 2006; Borra et al. 2008). More precisely, one pathway connects the upper bank of the STS with area PFG, while the other, originating from the lower bank of the STS, reaches areas F5a/p through AIP. Although these two studies did not compare action observation with action execution, they indicated further premotor and parietal areas where single neuron recording could reveal some new features for better elucidation of how cerebral cortex codes others' actions.

The Human Mirror System

The studies on monkeys clearly show, at the single neuron level, the existence of a parieto-frontal system matching observation and execution of motor acts. At the behavioral level it is obvious that the human species maintained the capacity of understanding actions of others. It is also very likely that the mirror neuron matching system was retained during evolution. However, for ethical reasons in humans, it is very difficult to give a direct demonstration of the existence of mirror neurons. In fact, intracortical recording is allowed only in epileptic patients in whom long electrodes are implanted for a limited period to better localize the epileptic foci. However, also in this case, these electrodes provide only electroencephalogram (EEG) data on the neural activity closer to each electrode. Actually, in one study, it was possible to record data from only single neurons (see below).

For these reasons, the mirror system in humans has been demonstrated with electrophysiological (transcranial magnetic stimulation [TMS], EEG, and magnetoencephalography [MEG]) and neuroimaging (positron emission tomography [PET], and fMRI) techniques that reveal the behavior not of single neurons but of populations of neurons. During observation or execution of motor acts, electrophysiological techniques allow a better temporal resolution of the activity correlated to these events, while the neuroimaging techniques, in particular fMRI, allow a better spatial resolution.

The first study trying to demonstrate the presence of an observation or execution matching system in humans was that by Fadiga et al. (1995), who stimulated the hand representation of the motor cortex of subjects observing an experimenter grasping an object. The idea underlying the study was that mirror neurons in the premotor cortex could specifically activate, during observation, the excitable motor cortex, but under threshold, and the TMS pulse should render this activation overt. Observation of meaningless arm movements, pure object observation and detection of a light dimming were the control conditions. During grasping observation with respect to control conditions, the magnetic stimulus given during observation showed a specific enhancement of the electromyographic activity (motor evoked potentials, MEPs) of those muscles that subjects normally use to execute the observed motor act. A further result of this study is that an enhanced activity was also found during observation of meaningless arm movements, suggesting that in humans, in contrast to monkeys, also this latter category of movements could activate corresponding motor representations. In a subsequent study based on a similar rationale, Gangitano et al. (2004) stimulated the excitable motor cortex of subjects with TMS observing reaching-grasping motor acts. The stimulation was given at different time points of the observed movements (e.g., before the hand started to move, during finger opening, during finger closure, etc.) and the corresponding MEPs were recorded. The results showed that the profile of cortical activation was in concordance with the timing of the kinematic profile of the observed finger movements. They concluded that the resonant motor plan retrieved during observation is loaded as a whole at the beginning of observation and once started tends to proceed to its completion regardless of changes in the visual cues.

In a recent study Cattaneo et al. (2009) used the TMS technique in subjects observing goal directed (grasping an object) and no-goal-directed (closure and opening without object)—hand movements performed with two different types of pliers, normal and reversed, as those used in the experiment of Umiltà et al. (2008) described above. TMS pulses were delivered over the hand representation of the left motor cortex and MEPs were recorded from opponens pollicis muscle, that is active during finger closure. The results showed that during the observation of no-goal movements MEPs were enhanced in relation with observed finger closure, independent of the type of pliers used, during observation of goal-directed movements MEPs were increased when both normal and inverted pliers grasped the object, that is during flexion in the first case and extension in the second case. In other words, it was the goal of the act and not the simple movement pattern that determined motor cortex activation.

The TMS technique, on one hand, provides a good demonstration that observed motor acts are matched with their motor representation and on the other hand gives only an approximate indication of the anatomical location of the activated areas. This issue was better addressed by several PET and fMRI studies, showing that action observation activated three main cortical sectors, mainly in the left hemisphere: one around the STS, a second one in the supramarginal gyrus (part of IPL), and a third one in the posterior sector of the inferior frontal gyrus (IFG). This latter sector corresponds to Brodmann's areas 44 and 45, forming the so-called Broca's area (the 'speech' area) (Rizzolatti et al. 1996b; Iacoboni et al. 1999; Buccino et al. 2001; Koski et al. 2002; Grèzes et al. 2003; Frey and Gerry 2006; see also Rizzolatti et al. 2009; Fogassi and Ferrari 2011). The IFG and IPL sectors correspond anatomically to the areas where mirror neurons have been described in monkeys (F5 and PFG, respectively). The STS sector corresponds to the area in the monkey described by Perrett et al. (1989), where neurons responding during observation of biological actions but devoid of motor properties have been recorded.

More recent fMRI studies showed that observation of grasping motor acts activates also the anterior intraparietal area (AIP) (Shmuelof and Zohari 2008) and that the activation of each hemisphere represents the identity of the observed contralateral hand. Furthermore, the activation of AIP is higher during observation of complex actions with respect to simple actions.

Most of neuroimaging studies revealed the activation of the above described fronto-parietal circuit. However, during observation of actions involving reaching movements or movements of several body parts, including the trunk and the leg, a more dorsal cortical sector, including dorsal premotor cortex and superior parietal lobule, could be activated (Calvo-Merino et al. 2005; Filimon et al. 2007). This suggests that the sector of activation depends on the observed acting effector. In this line, particularly interesting is the study by Buccino et al. (2001), showing that in subjects observing goal-related motor acts performed with the mouth (e.g., biting an apple), the hand (grasping a glass) and the leg (kicking a ball) there was a somatotopic activation of frontal and parietal cortices. Interestingly, when subjects observed the same motor acts, but pantomimed, the activation involved the same premotor and IFG regions activated by observation of goal-directed motor acts (Buccino et al. 2001; for similar findings see Grèzes et al. 1998), while parietal cortex was not activated. However, this could be due to a higher sensitivity of parietal cortex to the object target of the motor act. On the contrary, in another study (Lui et al. 2008), observation of symbolic gestures activated both ventral premotor and inferior parietal cortex, but this latter activation involves more posterior sectors than those normally activated by observation of goal-directed motor acts.

Another issue to be addressed is relative to the timing of cortical activation during action observation. The appropriate techniques for investigating this issue are electroencephalography (EEG) and magnetoencephalography (MEG). The use of these techniques confirmed that during both action observation and action execution there is an activation of motor cortex (Cochin et al. 1999) and that during observation, the activation of the IFG followed that of the visual cortex, and preceded that of primary motor cortex (Nishitani and Hari 2000).

As mentioned before, ethical reasons do not allow to provide a direct demonstration of the matching system at the level of single neurons. Although many neuroimaging studies involving both observation and execution of motor acts showed an overlap between areas activated during both conditions, one could argue that even when there is a perfect overlap, there could still be two different populations of neurons, one activated during observation and the other during execution (see for example, Dinstein et al. 2007). Some studies tried to assess the overlap between observation and execution activations, by applying a single-voxel analysis. This analysis demonstrated the activation of the same voxels in the two conditions in single subjects (see for example Gazzola et al. 2006).

In order to assess whether single neurons are activated in humans by both observation and execution of the same motor act, several authors used an indirect approach with the repetition-suppression fMRI technique (Grill-Spector et al. 2006). The rationale underlying this approach is that if the observation or execution matching occurs at the single neuron level, when subjects observe a motor act and then execute it or vice versa, the activation during the second event should be lower than that during the first event, as a result of an adaptation mechanism. This mechanism is well known from neurophysiological experiments demonstrating that several areas, in particular those of the ventral visual stream, adapt when the same *sensory* stimulus is repetitively presented (see for example Gross 1967; Ringo 1996; Kohn 2007; De Baene and Vogels 2010; Kaliukhovich and Vogels, 2011). While this phenomenon is likely to occur because of a decrease of activity at the presynaptic level, it cannot explain a motor-to-sensory adaptation or vice versa. Anyway, the attempts to demonstrate the existence of a mirror mechanism using fMRI adaptation techniques gave contradictory results (Dinstein et al. 2007; Chong et al. 2008; Kilner et al. 2009; Lingnau et al. 2009). Note, however, to claim that there could be two different populations in premotor areas, one responding only during execution and the other only during observation, would imply that there are purely visual neurons in the premotor cortex and that these latter ones do not communicate with motor neurons, an idea that is contradicted by single neuron data obtained from premotor cortex.

Recently, Mukamel et al. (2010) were able to record responses from single neurons of epileptic patients implanted with deep electrodes. During recording the patients were required to observe and execute reaching–grasping motor acts and facial expressions. They recorded from mesial cortical structures such as pre-supplementary (pre-SMA) and supplementary (SMA) motor areas, anterior cingulate cortex, hippocampus, parahippocampal cortex and entorhinal cortex. Neurons responding to both observation and execution were found in SMA and hippocampus. Interestingly, some neurons showed excitation during execution and inhibition during observation. As areas on the medial wall such as SMA seem relevant to movement initiation and movement sequences, they suggest that the neurons matching observation with execution recorded from SMA could be involved in these functions. However, neurons recorded in middle temporal cortex could represent the retrieval of memory of the action formed during action execution. It must be taken into account that in these patients, the placement of electrodes, based only on clinical considerations, do not allow to record from more lateral sectors of the cortex, those in which mirror neurons were typically recorded.

Sensitivity of the Mirror System to Observation of Tool-use

it has been demonstrated in monkeys, that after the training to use specific types of tools, some mirror neurons achieve the capacity to become active during observation of motor acts performed with tools (Umiltà et al. 2008; Rochat et al. 2010). It is plausible that in humans also, the mirror system could be sensitive to actions performed with tools by others, even though the real representation of the meaning of a tool is likely present only in humans and, may be at a lower degree, in apes. Gazzola et al. (2007a) instructed volunteers to observe videoclips of a human subject or a robot arm grasping objects; the results showed that the parieto-frontal mirror circuit was activated in both conditions. This finding was replicated and further extended by Peeters et al. (2009), who compared fMRI activation in monkeys and humans during the observation of motor acts performed by a human hand, a robot hand and various types of tools. Monkeys were scanned before and after learning tools. The aim was to see whether both monkeys and humans possess specific regions devoted to recognition of the tool actions. The results showed that, regardless of the type of effector used, the ventral premotor-inferior parietal circuit was always active in both humans and monkeys. However, during tool action observation there was a specific activation of a rostral sector of the left anterior supramarginal gyrus only in humans. This activation was not present in monkeys, even after a training to use the tools that were subsequently observed by the monkeys during fMRI acquisition. These findings imply important evolutionary considerations, because they suggest that the achievement of the capacity to use tool and to understand their meaning corresponded to the formation of a phylogenetically new cortical region.

A system matching sensory and motor representations, such as the mirror neuron system, appears to be very suitable to explain, at least in part, several social cognitive functions. For this reason, it has been proposed that it could constitute the basic scaffold for imitation, intention and emotion understanding, and for understanding of language. In the next sections we will describe evidence in favour of this proposal for some of these functions.

Mirroring and Imitation

The proposal that mirror neurons can be responsible for imitation was formulated several years ago (Jeannerod 1994; Rizzolatti et al. 2001). However, it is known that monkeys do not show "true imitation," that is, the capacity to immediately reproduce novel movements (Visalberghi and Fragaszy 1990). However, they show several types of imitative phenomena, such as, for example, neonatal imitation (imitation of mouth gestures in the first days of life, see Ferrari et al. 2006) or action facilitation (Ferrari et al. 2005). All these behaviors imply the translation of the behavior of the observed agent into motor programs necessary to reproduce the observed movements that is the retrieval of corresponding motor representations.

Basic facial gestures such as lipsmacking and tongue protrusion in response to the same human facial gestures (neonatal imitation) has been recently described in infant macaque monkeys (Ferrari et al. 2006). On the basis of these results, it was concluded that neonatal imitation and imitative abilities are not unique to apes and humans. The authors proposed that in both humans and monkeys a primitive mirror mechanism can be present at birth and this would be responsible for matching the facial features with the internal corresponding motor representation. In a second preliminary EEG study (Ferrari et al. 2008), the authors reported an alpha rhythm suppression in one-week-old infant macaques when they were observing facial gestures but not during the observation of non-biological movements. As this decrease in activity has been considered to reflect the activation of areas recorded in the central-parietal motor regionin humans, this result has been considered as evidence of the activation of the mirror system.

Imitation in Humans

In humans, the interest in imitation studies has increased more and more in recent years, thanks to the possibility of studying this function with neuroimaging and high resolution electrophysiological techniques. A series of fMRI studies by the group of Iacoboni used a paradigm in which subjects had to observe and imitate simple finger movements (Iacoboni et al. 1999). The results showed an activation of the IFG and of the inferior parietal cortex during imitation. Moreover, they found an activation of the superior parietal cortex, attributed to the activation of kinesthetic hand representations, that are typically found, at the single neuron level, in this part of the cortex. Subsequent works of the same group basically replicated these findings when in the same paradigm a new goal was introduced which was to touch a spot on the table with the finger. (Koski et al. 2002). In addition, the activity was higher when target-directed movements were compared with those without a goal.

The most important aspect of imitation in humans is imitation learning, that is, the capacity to learn new skills by observation. This is the mean through which we learn to play music instruments, sports, dance and many types of complex motor tasks. Imitation learning is based on the observation of what another agent is doing, from a third person or an egocentric perspective, followed by subsequent attempts to reproduce the model. Although the observed behavior can be quite complex, it is normally constituted by sequences of motor acts. As we know, motor acts are understood through mirror neuron activity. Therefore it is plausible that in a task of imitation learning the mirror system is active during both observation and imitation.

This issue was tested by Buccino et al. (2004), in an event-related fMRI study in which participants had to observe and then, after a pause, imitate novel guitar chords played by an expert guitarist. The study showed that, there was a strong activation of the IPL and ventral premotor cortex plus the pars opercularis of IFG during observation and even more so during the imitation phase. During imitation, of course, there was also a strong activation of primary motor and somatosensory cortices due to the actual execution of the chords. Thus, these data confirm the strong involvement of the mirror system in imitation learning. However, this system is not enough for this function. In fact, the same study shows that during the phase interleaved between observation and imitation, when subjects were preparing the program for correctly reproducing the observed chord, there was an additional strong activation of the middle prefrontal cortex (BA 46). This indicates the possible mechanism occurring during imitation. Mirror neurons are responsible for matching observed motor acts with their motor representation in the observer, enabling the observer to automatically understand these acts. These acts, however, must be organized in the same sequence used by the model. This latter task would require the activation of prefrontal cortex, a cortical sector known for its role in action planning. As a matter of fact, it activates during the pause phase, but it becomes silent during the actual imitation.

Mirror Neurons and Intention Coding

The issue of intention coding is very relevant in humans because it relates to our capacity of understanding others' intentions, a component of mindreading. There is a large debate about the mechanisms underlying this function. Theory-theory, which was especially used for explaining developmental data, postulates that intention attribution emerges from a theoretical reasoning that assumes the existence of laws linking external stimuli to internal states, and these latter to behavior. By virtue of these laws, intention understanding would be achieved through an inferential process. On the contrary, according to the simulation theory, the observation of the behavior of another individual would determine an internal simulation of her/his actions. Thus, while theory-theory requires reasoning, a time-consuming process, simulation theory implies an automatic retrieval of motor representations, which is in a very fast process. The data described in monkeys on action organization and intention coding seem closer to the assumptions of the simulation theory.

Most of the neurons described in both ventral premotor and parietal cortex are involved in encoding the goal of motor acts. However, as motor acts are put in sequences to form actions, it is important to address the issue of whether, at some level, actions also are encoded. A first answer to this question came from series of experiments (Fogassi et al. 2005; Bonini et al. 2010, 2011) carried out in order to evaluate whether the discharge of grasping neurons can be influenced by the type of action in which the grasping act is embedded. The activity of grasping neurons located in the inferior parietal area PFG and in ventral premotor area F5 were recorded while the monkey performed a motor task and observed the same task executed by an experimenter. The task consisted of two conditions in which the same grasping act was executed in order to attain two different goals (eating or placing). The first part of the tasks was identical in the two conditions, i.e., the monkey or experimenter starting from a fixed position grasped a piece of food. Then the monkey or experimenter had to bring the food to the mouth in order to eat it (grasp to eat) or had to place it in a container positioned near the mouth or near the target (grasp to place). The results showed that even though the grasping acts executed and observed by the monkeys were identical in the eating and placing conditions, both parietal and premotor grasping neurons discharged differently according to the goal of the action in which the grasping act was embedded. This finding indicates that the final action goal (motor intention) can modulate the discharge of neurons encoding motor acts. This modulation is probably used to activate the neurons coding the subsequent motor act of the sequence directly involved in achieving the final goal. For example, in the eating condition, the grasping act precedes the act of bringing food to the mouth, while in the placing condition, it precedes the act of reaching a container. On the basis of these results it has been hypothesized that neurons belonging to the parieto-premotor circuit are organized in prewired intentional chains in which each neuron coding a motor act facilitates the neuron coding the subsequent motor act.

During the motor task, since the monkey knows its intention, the differential discharge of grasping neurons appears to *reflect* the monkey motor intention. During the visual task, since the grasping act is executed by the observed agent, the differential visual response appears to *predict* the action outcome. The mechanism should be as follows: the observation of the first motor acts of the action sequence within a certain context, retrieves a specific motor chain coding a given intention, thus allowing to automatically understand the intention of the observed agent. The activation of specific neuronal chains, each corresponding to an action goal, could constitute the primitive mechanism upon which more sophisticated mindreading mechanism could have been built.

The possibility that a mirror mechanism can play a role also in understanding others' intentions in humans was first suggested by an fMRI study by Iacoboni et al. (2005). They asked subjects to observe three conditions: a videoclip that could represent either a breakfast to be started or finished (context condition), a videoclip showing a hand grasping a cup in a empty background (action condition) and a hand grasping a cup in one of the contexts of the first condition (intention condition). The subjects were divided into two groups: subjects of the first group were instructed to simply watch the movies, those of the second group were told to understand the intention underlying the grasping action within each context. The most important result was that the intentioncondition (aimed to reveal the specific effect of the intention underlying the action) produced a stronger activation, relative to the other two conditions, of the right IFG. Interestingly, this effect was independent of whether the subject were asked to purely observe the three types of videoclips or to observe in order to explicitly understand the intention.

The studies reviewed above indicate that the parieto-frontal mirror network subserves the understanding of motor intentions underlying the actions of others. This does not mean that the parieto-frontal mirror mechanism covers all types of intention understanding. While motor intention can be automatically understood through a process of retrieval of action representations, the interpretation of others' behavior can, in some situations, require reasoning, thus involving further cortical areas, such as those considered to be part of the "mentalizing" network (see Brass et al. 2007; de Lange et al. 2008; Liepelt et al. 2008). Brass et al. (2007) addressed this issue with an fMRI experiment in which subjects had to observe several videoclips showing plausible and unplausible actions. For example, a videoclip showed a person turning on a light by pressing a switch with a knee, in a condition in which she carried out two folders with the hands (plausible-constraint) and in another condition in which the hands were free (no-constraint). The comparison of the activation between the two situations did not reveal any differential activation of the mirror system, while a reliable activation of the STS region and a little less reliable activation of the anterior fronto-median cortex was found. Both regions belong to the "mentalizing" network, that also includes the temporo-parietal junction (TPJ) and the posterior cingulated cortex. Thus, in the case in which it is necessary to make inferences in order to understand the reasons underlying one's behavior, other areas besides the mirror system can come into play. Note, however, that while the mirror neuron matching mechanism is known in depth, thanks to single neuron studies in monkeys, the neurophysiological mechanisms underlying mentalizing, beyond the description of the activated areas, are unknown.

The topic of intention understanding captured great attention in the field of the disorders of social interactions. It is well known that at the core of the autistic syndrome there is an impairment in social interactions, in which the difficulties in understanding others' intentions and emotions and those in imitating others' behavior are crucial symptoms. Because of this, many authors have already proposed that a malfunctioning mirror system could explain these social cognitive impairments (Williams et al. 2001; Ramachandran and Oberman 2006). Several findings go in this direction. For example it has been shown that during observation of motor acts autistic patients do not show the typical mu rhythm desynchronization of the central cortex shown by typically developed individuals (Oberman et al. 2005). Furthermore, they show a lower activation of the IFG during observation and imitation of emotional expressions (Dapretto et al. 2006). The regions included in the mirror system appear to be thinner in the gray matter with respect to other cortical regions (Hadjikhani et al. 2006).

However, it must be considered that among the impairments ascribed to the autistic syndrome there are motor deficits. These disturbances consist in several deficits such as clumsiness, postural instability, disturbances in motor coordination.

In order to understand whether action organization is normal in high functioning autistic children, Cattaneo et al. (2007) tested both autistic (ASD) and typically developed children (TD) in a task in which subjects had to perform two actions similar to those used in the single neuron study by Fogassi et al. (2005), grasping for eating and grasping for placing. During the performance of this task, they recorded the EMG of the mylohyoid muscle, which is involved in mouth closure, observing its activation during the two actions. They found that while in TD this muscle activates several hundred milliseconds before the subject's hand touches the object to be grasped, in ASD this activation began after this event. Since the early activation in normal behavior is due to the programming of a motor act already during the preceding motor act, in order to make the action fluent, the lack of this early activation in ASD children suggest a deficit in the organization of intentional actions. In agreement with this, when the two groups of subjects had to observe the same actions performed by an experimenter, while TD children showed the mylohyoid muscle activation also during observation, with the same timing of execution, the EMG of ASD children was flat. Thus, it appears that the impairment shown during observation depends on the basic malfunctioning of the intentional encoding within the cortical motor system.

Learning-Induced Modulation of the Mirror System

The above described studies on imitation learning (Buccino et al. 2004), in which the mirror system appears to be more active during imitation, suggest that this system could show plastic properties.

In single neuron studies, in order to show that mirror neurons have this type of capacity, chronic recordings during a learning task are needed. Until now, this type of experiment has not been reported. However, the evidence for the presence of F5 mirror neurons responding to the observation of grasping motor acts performed by an experimenter with a tool (Ferrari et al. 2005; Umiltà et al. 2008; Rochat et al. 2010) is an indirect demonstration of a possible plasticity. The results of these studies point to the basic organization of the motor system as a scaffold for the matching of new types of goal-related motor acts. Is this principle crucial also in humans?

In humans, fMRI studies show that motor experience influence the activation of the motor system during action observation. In one study (Calvo-Merino et al. 2005) participants were classic ballet dancers of Capoeira (a latino-american dance, derived from a martial art dancers) and people naïve in professional dance. All three groups were instructed to observed videoclips showing steps of classical dance or Capoeira. All groups had an activation of the mirror system, but this was higher in the experts groups. The activation involved both parietal cortex and premotor cortex. More interestingly, it has been found that in Capoeira dancers the observation of Capoeira dance caused a greater activation with respect to observation of classical dance, while the opposite was observed in classical dancers during observation of classic ballet. Naïve subjects did not show any differential activation between the two conditions. These results suggest that, since the motor experience likely determines a modulation of the motor circuits related to that particular skill, during observation the motor "resonance" with the actions in which one is more skilled is higher than during observation of non practiced actions. This is also confirmed by the lack of differential activation observed in people who are not skilled in any of the observed expertise.

The plasticity of the mirror system is further demonstrated by studies in which fMRI scanning could be carried out several times, in parallel with learning of new dance sequences (Cross et al. 2006). Dancers had to observe and imagine performing movement sequences, half of which were rehearsed and half unpracticed. The results showed that a differential increase of activity in PMv and IPL during observation of the rehearsed sequences.

If the mirror system strongly relies on the observer's motor repertoire, how is its activation during situations in which individuals cannot have normal motor experience? Gazzola et al. (2007b) addressed this issue in an fMRI study in which subjects, two aplasic individuals born without arms or hands, and TD individuals had to observe goal-related hand motor acts and to execute mouth, hand (only TD) and foot motor acts. Surprisingly, the study showed that aplasic subjects had an activation of the mirror system not different from TD subjects. An even more surprising finding was that the activation of aplasic subjects during grasping observation corresponded anatomically to the frontal sectors activated during execution of mouth and foot actions. This suggest that, in absence of a hand motor representation, hand action observation retrieved those cortical motor representations involved in the execution of motor acts that achieve similar goals (e.g., taking possession of an object) with effectors different from the hand. These results are in support of a plasticity of the mirror system since neonatal life.

Previous studies aimed to exploit the plasticity of the motor system, used motor training or motor imagery as tools in rehabilitation. On the basis of the plastic features of the mirror system, the interest is now growing for the possible exploitation of the observation or execution system for rehabilitative purposes. Recently, action observation therapy has been employed to assess whether action observation could improve the motor performance of patients with paresis and whether this therapy could modify brain activity. Ertelt et al. (2007) employed this therapy on stroke patients presenting a mild paretic hand. The patients were subdivided into two groups: one (experimental group) was instructed to observe and reproduce motor acts of increasing complexity, and the other (control group) to observe videos showing geometric symbols and letters, and then to perform the same motor acts as the first group. The therapy lasted for 18 days. After the end of therapy, the data showed, only for the experimental group, an improvement of the functional scales used to evaluate motor performance. In order to see whether the therapy induced objective changes in brain organization, part of the patients of both experimental and control groups underwent two fMRI scans, one before and the other after rehabilitation, in which they had to perform an object manipulation task. The study revealed an increased activation, in the experimental group as compared with that in the control group, in areas of both hemispheres, many of which (PMv, SMG and STS) belonged to the observation or execution system.

A recent pilot study (Da Silva Cameirao et al. 2011) used a self-managed rehabilitation approach, in which a group of acute stroke patients used a rehabilitation gaming system (RGS), while the control group performed a time matched alternative treatment, consisting of an intense occupational therapy or non-specific interactive games. The RGS required the execution and observation (through virtual reality) of motor acts such as hitting, grasping or placing a spherical object. The therapy lasted for 12 weeks. At the end of the therapy, the RGS group showed a significantly improved performance in paretic arm speed as demonstrated by tests specific for the forelimb and all clinical scales evaluating functional daily life activities.

Evolutionary Aspects of the Mirror Mechanism

The Mirror Matching Mechanism and Language Evolution

One of the theories challenging the view that human language has appeared as a full-fledged function, postulates that gestures are at the origin of spoken language (Corballis 2002). This is confirmed by the still existing coupling between gestures and utterances during inter-individual communication or by the use of sign language in deaf persons. For language comprehension to occur, the sender and the receiver must share the same neural substrate. This is one of the reasons why the mirror matching mechanism has been considered as a possible candidate for the evolution of this type of interaction. Although in monkeys, mirror neurons are mostly related to hand or mouth goal-related motor acts, the presence in area F5 of audio-visual mirror neurons (Kohler et al. 2002) and of communicative mirror neurons (Ferrari et al. 2003) suggest that in this area there can be some prototypes of speech production and understanding (Rizzolatti and Arbib 1998; Fogassi and Ferrari 2007). Furthermore, area 44 (the posterior sector of Broca's area) is considered anatomically homologue to part of area F5 (see Rizzolatti and Arbib 1998; Petrides et al. 2005) and a functional comparison shows that the two areas share many properties. In fact, they both contain hand and mouth high order representation, are activated by observation of hand and mouth actions and by listening to action sounds. Furthermore, Broca's area is active during imitation. How, however, was action understanding translated in speech understanding? A possible important step could have been represented by the understanding of intransitive gestures endowed with a symbolic meaning (see Arbib 2005). Another important evolutionary achievement is also represented by the combination of meaningful utterances and gestures, which was demonstrated in chimpanzees. As mentioned above, this combination is still present in humans (McNeill 2000; Gentilucci and Corballis 2006). Later in evolution the acquisition of a more sophisticated articulatory apparatus led vocal production to dominate on brachiomanual communication, one of the reasons likely being a better adaptation to situations in which visual interaction was not possible.

Many investigations of the last decade demonstrated how the motor system is influenced during listening to verbal material. A description of the studies on this topic goes beyond the aim of this article. However, it is interesting to mention a few human studies demonstrating motor resonance during speech listening.

A TMS study (Fadiga et al. 2002) showed that when subjects listened to words that, when executed, require a strong involvement of tongue muscles, there was a higher MEP activation of these muscles, as compared to the condition in which subjects listened to words normally requiring a lower muscle involvement. Thus, during language perception, it is very likely that there is a phonological resonance of the motor representations involved in producing the same listened material. These data seem to be in agreement with Liberman's motor theory of speech perception (Liberman and Mattingly 1985), proposing that this function is not based on an acoustic analysis, but on the capacity of sharing motor invariants (of the utterances) between the sender and the receiver. Although these data do not explain completely the presence of an intact speech perception associated to an impaired speech production, they suggest a very important mechanism through which children could understand phonology, thus learning new words.

Other studies indicate that the motor system can be involved in understanding action verbs. Hauk et al. (2004), using fMRI, showed that reading action-related words, such as lick, pick and kick, differentially activated areas of the motor cortex that were directly adjacent to or overlapped with that activated during execution of movements of the tongue, fingers, or feet. They concluded that this activation rules out the existence of a unique meaning center. In a similar vein, Tettamanti et al. (2005) showed that listening to action-related sentences (e.g., I grasp the glass) as compared to abstract sentences (I love justice) differentially activate sectors of premotor cortex. Also, in this case, as in the previous study, the activation was somatotopically organized. This evidence indicates that a possible mirror mechanism can be responsible for the motor resonance during perception of motorically meaningful verbal material. If this is accepted, one of the issues to be investigated in the future is whether there are two different, but adjacent, mirror systems, one for action understanding and the other for language comprehension, the latter being phylogenetically derived from the former.

Mirror Neurons for Songs in Songbirds

In evolutionary terms, the mirror neuron matching mechanism could be a quite old achievement. According to this idea, very recently, a matching process similar to those illustrated in humans and monkeys has been described in birds (Prather et al. 2008). The authors reported a class of auditory-motor neurons, recorded in sparrow forebrain, that were active during both listening to and singing the species-specific song. According to the authors' interpretation, this neural substrate would facilitate vocal communication in two ways. First, the acoustic activation of the receiver's auditory–vocal neurons, evoked by sender's vocalizations, allows the comparison between the heard vocalizations and the internal representation of the receiver's vocal motions. Second, acoustic activation of the auditory–vocal neurons by other birds' songs could furnishes a template that can be used by the animal to select a song that matches with that of its tutor, facilitating the learning processes. Thus, even in the case of mirror neurons for songs, the mechanism relies on the convergence at single neuron level of sensory and motor information used to represent a signal.

The presence of a perception or production matching mechanism in songbirds indicates that action-perception matching mechanisms are probably parsimonious solutions the brain of vertebrate has evolved for processing complex biological stimuli. Recently, has been proposed an homology between the avian pallial and mammalian cerebral cortex. Thus, it is possible that some primate areas involved in motor-cognitive functions are not so different from avian brain structures endowed with similar mechanisms. Within this new framework, the matching mechanism demonstrated in songbirds would share with primates a common phylogenetic origin, probably based on a similar motor scaffold, capable of connecting motor representations with biological input.

References

Arbib M (2005) From monkey-like action recognition to human language: an evolutionary framework for neurolinguistics. Behav Brain Sci 28:105–167

Baron-Cohen S (1994) How to build a baby that can read minds: cognitive mechanisms in mindreading. Curr Psychol Cogn 13:513–552

Barraclough NE, Xiao D, Oram MW, Perrett DI (2006) The sensitivity of primate STS neurons to walking sequences and to the degree of articulation in static images. Prog Brain Res 154:135–148

Bonini L, Rozzi S, Ugolotti Serventi F, Simone L, Ferrari PF, Fogassi L (2010) Ventral premotor and inferior parietal cortices make distinct contribution to action organization and intention understanding. Cerebr Cort 20:1372–1385

Bonini L, Ugolotti Serventi F, Simone L, Rozzi S, Ferrari PF, Fogassi L (2011) Grasping neurons of monkey parietal and premotor cortices encode action goals at distinct levels of abstraction during complex action sequences. J Neurosci 31:5876–5887

Borra E, Belmalih A, Calzavara R, Gerbella M, Murata A, Rozzi S, Luppino G (2008) Cortical connections of the macaque anterior intraparietal (AIP) area. Cereb Cortex 18:1094–1111

Brass M, Schmitt RM, Spengler S, Gergely G (2007) Investigating action understanding: inferential processes versus action simulation. Curr Biol 17:2117–2121

Buccino G, Binkofski F, Fink GR, Fadiga L, Fogassi L, Gallese V, Seitz RJ, Zilles K, Rizzolatti G, Freund HJ (2001) Action observation activates premotor and parietal areas in a somatotopic manner: an fMRI study. Eur J Neurosci 13:400–404

Buccino G, Vogt S, Ritzl A, Fink GR, Zilles K, Freund HJ, Rizzolatti G (2004) Neural circuits underlying imitation of hand actions: an event related fMRI study. Neuron 42:323–334

Caggiano V, Fogassi L, Rizzolatti G, Their P, Casile A (2009) Mirror neurons differentially encode the peripersonal and extrapersonal space of monkeys. Science 324:403–406

Caggiano V, Fogassi L, Rizzolatti G, Pomper JK, Their P, Giese MA, Casile A (2011) View-based encoding of actions in mirror neurons of area f5 in macaque premotor cortex. Curr Biol 21:144–148

Calvo-Merino B, Glaser DE, Grezes J, Passingham RE, Haggard P (2005) Action observation and acquired motor skills: an FMRI study with expert dancers. Cereb Cortex 15:1243–1249

Cattaneo L, Fabbri-Destro M, Boria S, Pieraccini C, Monti A, Cossu G, Rizzolatti G (2007) Impairment of actions chains in autism and its possible role in intention understanding. Proc Natl Acad Sci U S A 104:17825–17830

Cattaneo L, Caruana F, Jezzini, A, Rizzolatti G (2009) Representation of goal and movements without overt motor behavior in the human motor cortex: a transcranial magnetic stimulation study. J Neurosci 29:11134–11138

Cisek P, Kalaska JF (2005) Neural correlates of reaching decisions in dorsal premotor cortex: specification of multiple direction choices and final selection of action. Neuron 45:801–814

Chong TT, Cunnington R, Williams MA, Kanwisher N, Mattingley JB (2008) fMRI adaptation reveals mirror neurons in human inferior parietal cortex. Curr Biol 18(20):1576–1580

Cochin S, Barthelemy C, Roux S, Martineau, J (1999) Observation and execution of movement: similarities demonstrated by quantified electroencephalograpy. Eur J Neurosci 11:1839–1842

Corballis MC (2002) From hand to mouth: the origins of language. Princeton University Press, Princeton

Cross ES, de Hamilton AF, Grafton ST (2006) Building a motor simulation de novo: observation of dance by dancers. NeuroImage 31:1257–1267

Dapretto M, Davies MS, Pfeifer JH, Scott AA, Sigman M, Bookheimer SY, Iacoboni M (2006) Understanding emotions in others: mirror neuron dysfunction in children with autism spectrum disorders. Nat Neurosci 9:28–30

De Baene W, Vogels R (2010) Effects of adaptation on the stimulus selectivity of macaque inferior temporal spiking activity and local field potentials. Cereb Cortex 20:2145–2165

Da Silva Cameirão M, Bermúdez i Badia S, Duarte E, Verschure PF (2011) Virtual reality based rehabilitation speeds up functional recovery of the upper extremities after stroke: a randomized controlled pilot study in the acute phase of stroke using the Rehabilitation Game System. Restor Neurol Neurosci 29:287–298

de Lange FP, Spronk M, Willems RM, Toni I, Bekkering H (2008) Complementary systems for understanding action intentions. Curr Biol 18:454–457

Dinstein I, Hasson U, Rubin N, Heeger DJ (2007) Brain areas selective for both observed and executed movements. J neurophysiol 98:1415–1427

Di Pellegrino G, Fadiga L, Fogassi L, Gallese V, Rizzolatti G (1992) Understanding motor events: a neurophysiological study. Exp Brain Res 91:176–180

Ertelt D, Small S, Solodkin A, Dettmers C, McNamara A, Binkofski F, Buccino G (2007) Action observation has a positive impact on rehabilitation of motor deficits after stroke. NeuroImage 36 (Suppl 2):T164–173

Evarts EV (1968) Relation of pyramidal tract activity to force exerted during voluntary movement. J Neurophysiol 31:14–27

Fadiga L, Fogassi L, Pavesi G, Rizzolatti G (1995) Motor facilitation during action observation: a magnetic stimulation study. J Neurophysiol 73:2608–2611

Fadiga L, Craighero L, Buccino G, Rizzolatti G (2002) Speech listening specifically modulates the excitability of tongue muscles: a TMS study. Eur J Neurosci 15:399–402

Ferrari PF, Rozzi S, Fogassi L (2005) Mirror neurons responding to observation of actions made with tools in monkey ventral premotor cortex. J Cogn Neurosci 17:212–226

Ferrari PF, Gallese V, Rizzolatti G, Fogassi L (2003) Mirror neurons responding to the observation of ingestive and communicative mouth actions in the monkey ventral premotor cortex. Eur J Neurosci 17 1703–1714

Ferrari PF, Vanderwert R, Herman K, Paukner A, Fox NA, Suomi SJ (2008) EEG activity in response to facial gestures in 1–7 days old infant rhesus macaques. Soc Neurosci Abs 297:13

Ferrari PF, Visalberghi E, Paukner A, Fogassi L, Ruggiero A, Suomi SJ (2006) Neonatal imitation in rhesus macaques. PLoS Biol 4:e302

Filimon F, Nelson JD, Hagler DJ, Sereno MI (2007) Human cortical representations for reaching: mirror neurons for execution, observation, and imagery. NeuroImage 37:1315–1328

Fogassi L, Ferrari PF (2007) Mirror neurons and the evolution of embodied language. Curr Dir Psych Sci 16:136–141

Fogassi L, Ferrari PF (2011) Mirror systems. WIREs Cogn Sci 2:22–38. doi:10.1002/wcs.89

Fogassi L, Ferrari PF, Gesierich B, Rozzi S, Chersi F, Rizzolatti G (2005) Parietal Lobe: from Action Organization to Intention Understanding. Science 308:662–667

Frey SH, Gerry VE (2006) Modulation of neural activity during observational learning of actions and their sequential orders. J Neurosci 26:13194–13201

Gallese V, Fadiga L, Fogassi L, Rizzolatti G (1996) Action recognition in the premotor cortex. Brain 119:593–603

Gallese V, Fadiga L, Fogassi L, Rizzolatti G (2002) Action representation and the inferior parietal lobule. In: Prinz W, Hommel B (eds) Common mechanisms in perception and action: attention and performance. Oxford University Press, Oxford, pp 334–355

Gangitano M, Mottaghy FM, Pascual-Leone A (2004) Modulation of premotor mirror neuron activity during observation of unpredictable grasping movements. Eur J Neurosci 20:2193–2202

Gazzola V, Aziz-Zadeh L, Keysers C (2006) Empathy and the somatotopic auditory mirror system in humans. Curr Biol 16:1824–1829

Gazzola V, Rizzolatti G, Wicker B, Keysers C (2007a) The anthropomorphic brain: the mirror neuron system responds to human and robotic actions. NeuroImage 35:1674–1684

Gazzola V, Van Der Worp H, Mulder T, Wicker B, Rizzolatti G, Keysers C (2007b) Aplasics born without hands mirror the goal of hand actions with their feet. Curr Biol 17:1235–1240
Gentilucci M, Corballis MC (2006) From manual gesture to speech: a gradual transition. Neurosci Biobehav Rev 30:949–960
Grèzes J, Costes N, Decety J (1998) Top–down effect of strategy on the perception of human biological motion: a PET investigation. Cogn Neuropsychol 15:553–582
Grèzes J, Armony JL, Rowe J, Passingham RE (2003) Activations related to "mirror" and "canonical" neurones in the human brain: an fMRI study. NeuroImage 18:928–937
Gregoriou GG, Borra E, Matelli M, Luppino G (2006) Architectonic organization of the inferior parietal convexity of the macaque monkey. J Comp Neurol 496:422–451
Grill-Spector K, Henson R, Martin A (2006) Repetition and the brain: neural models of stimulus-specific effects. Trends Cogn Sci 10:14–23
Gross CG, Schiller PH, Wells C, Gerstein GL (1967) Single-unit activity in temporal association cortex of the monkey. J Neurophysiol 30:833–843
Hadjikhani N, Joseph RM, Snyder J, Tager-Flusberg H (2006) Anatomical differences in the mirror neuron system and social cognition network in autism. Cereb Cortex 16:1276–1282
Hauk O, Johnsrude I, Pulvermüller F (2004) Somatotopic representation of action words in human motor and premotor cortex. Neuron 41:301–307
Iacoboni M, Woods RP, Brass M, Bekkering H, Mazziotta JC, Rizzolatti G(1999) Cortical mechanisms of human imitation. Science 286:2526–2528
Iacoboni M, Molnar-Szakacs I, Gallese V, Buccino G, Mazziotta JC, Rizzolatti G (2005) Grasping the intentions of others with one's own mirror neuron system. PLoS Biol 3:e79
Jeannerod M (1988) The neural and behavioural organization of goal-directed movements. Oxford University Press, Oxford
Jeannerod M (1994) The representing brain: neural correlates of motor intention and imagery. Behav Brain Sci 17:187–245
Kaliukhovich DA, Vogels R (2011) Stimulus repetition probability does not affect repetition suppression in macaque inferior temporal cortex. Cereb Cortex 21:1547–1558
Kilner JM, Neal A, Weiskopf N, Friston KJ, Frith CD (2009) Evidence of mirror neurons in human inferior frontal gyrus. J Neurosci 29:10153–10159
Kohler E, Keysers C, Umiltà MA, Fogassi L, Gallese V, Rizzolatti G (2002) Hearing sounds, understanding actions: action representation in mirror neurons. Science 297:846–848
Kohn A (2007) Visual adaptation: physiology, mechanisms, and functional benefits. J Neurophysiol 97:3155–3164
Koski L, Wohlschläger A, Bekkering H, Woods RP, Dubeau MC, Mazziotta JC, Iacoboni M (2002) Modulation of motor and premotor activity during imitation of target-directed actions. Cereb Cortex 12:847–855
Kraskov A, Dancause N, Quallo MM, Shepherd S, Lemon RN (2009) Corticospinal neurons in macaque ventral premotor cortex with mirror properties: a Potential mechanism for action suppression? Neuron 64:922–930
Liberman AM, Mattingly IG (1985) The motor theory of speech perception revised. Cognition 21:1–36
Liepelt R, Von Cramon DY, Brass M (2008) How do we infer other's goals from non stereotypic actions? The outcome of context-sensitive inferential processing in right inferior parietal and posterior temporal cortex. NeuroImage 43:784–792
Lingnau A, Gesierich B, Caramazza A (2009) Asymmetric fMRI adaptation reveals no evidence for mirror neurons in humans. Proc Natl Acad Sci U S A 106:9925–9930
Lui F, Buccino G, Duzzi D, Benuzzi F, Crisi G, Baraldi P, Nichelli P, Porro C, Rizzolatti G (2008) Neural substrates for observing and imagining non-object-directed actions. In: Keysers C, Fadiga L (eds) The mirror neuron system. Special Issue. Soc Neurosci Psychol Press, Hove, pp 261–275
McNeill D (2000) Language and gesture. Cambridge University Press, Cambridge
Mukamel R, Ekstrom AD, Kaplan J, Iacoboni M, Fried I (2010) Single-neuron responses in humans during observation and execution of actions. Curr Biol 20:1–7

Nelissen K, Luppino G, Vanduffel W, Rizzolatti G, Orban GA (2005) Observing others: multiple action representation in the frontal lobe. Science 310:332–336

Nelissen K, Borra E, Gerbella M, Rozzi S, Luppino G, Vanduffel W, Rizzolatti G, Orban GA (2011) Action observation circuits in the macaque monkey cortex. J Neurosci 31:3743–3756

Nishitani N, Hari R (2000) Temporal dynamics of cortical representation for action. Proc Natl Acad Sci U S A 97:913–918

Oberman LM, Hubbard EM, McCleery JP, Altschuler EL, Ramachandran VS, Pineda JA (2005) EEG evidence for mirror neuron dysfunction in autism spectrum disorders. Brain Res Cogn Brain Res 24:190–198

Pandya DN, Seltzer B (1982) Intrinsic connections and architectonics of posterior parietal cortex in the rhesus monkey. J Comp Neurol 204:196–210

Peeters R, Simone L, Nelissen K, Fabbri-Destro M, Vanduffel W, Rizzolatti G, Orban GA (2009) The representation of tool use in humans and monkeys: common and unique human features. J Neurosci 29:11523–11539

Perrett DI, Harries MH, Bevan R, Thomas S, Benson PJ, Mistlin AJ, Chitty AK, Hietanen JK, Ortega JE (1989) Frameworks of analysis for the neural representation of animate objects and actions. J Exp Biol 146:87–113

Petrides M, Cadoret G, Mackey S (2005) Orofacial somatomotor responses in the macaque monkey homologue of Broca's area. Nature 435:1235–1238

Prather JF, Peters S, Nowicki S, Mooney R (2008) Precise auditory-vocal mirroring in neurons for learned vocal communication. Nature 451:305–310

Ramachandran VS, Oberman LM (2006) Broken mirrors: a theory of autism. Sci Am 295:62–69

Ringo JL (1996) Stimulus specific adaptation in inferior temporal and medial temporal cortex of the monkey. Behav Brain Res 76:191–197

Rizzolatti G, Fadiga L, Gallese V, Fogassi L (1996a) Premotor cortex and the recognition of motor actions. Brain Res Cogn Brain Res 3:131–141

Rizzolatti G, Fadiga L, Matelli M, Bettinardi V, Paulesu E, Perani D, Fazio F (1996b) Localization of grasp representations in humans by PET: 1. Observation versus execution. Exp Brain Res 111:246–252

Rizzolatti G, Fogassi L, Gallese V (2009) The mirror neuron system: a motor-based mechanism for action and intention understanding. In: Gazzaniga M (ed) The cognitive neuroscience IV. The MIT Press, Cambridge, pp 625–640

Rizzolatti G, Fogassi L, Gallese V (1997) Parietal cortex: from sight to action. Curr Op Neurobiol 7:562–567

Rizzolatti G, Luppino G (2001) The cortical motor system. Neuron 31:889–901

Rizzolatti G, Arbib MA (1998) Language within our grasp. Trends Neurosci 21:188–194

Rizzolatti G, Fogassi L, Gallese V (2001) Neurophysiological mechanisms underlying the understanding and imitation of action. Nat Rev Neurosci 2:661–670

Rochat MJ, Caruana F, Jezzini A, Escola L, Intskirveli I, Grammont F, Gallese V, Rizzolatti G, Umiltà MA (2010) Responses of mirror neurons in area F5 to hand and tool grasping observation. Exp Brain Res 204:605–616

Rosenbaum DA, Cohen RG, Jax SA, Weiss DJ, Van Der Wel R (2007) The problem of serial order in behavior: Lashley's legacy. Hum Mov Sci 26:525–554

Rozzi S, Calzavara R, Belmalih A, Borra E, Gregoriou GG, Matelli M, Luppino G (2006) Cortical connections of the inferior parietal cortical convexity of the macaque monkey. Cereb Cortex 16:1389–1417

Rozzi S, Ferrari PF, Bonini L, Rizzolatti G, Fogassi L (2008) Functional organization of inferior parietal lobule convexity in the macaque monkey: electrophysiological characterization of motor, sensory and mirror responses and their correlation with cytoarchitectonic areas. Eur J Neurosci 28:1569–1588

Shepherd SV, Klein JT, Deaner RO, Platt ML (2009) Mirroring of attention by neurons in macaque parietal cortex. Proc Natl Acad Sci U S A 106:9489–9494

Shmuelof L, Zohary E (2008) A mirror representation of others' actions in the human anterior parietal cortex. Nat Neurosci 11:1267–1269

Tettamanti M, Buccino G, Saccuman MC (2005) Listening to action-related sentences activates fronto-parietal motor circuits. J Cogn Neurosci 17:273–281

Tkach D, Reimer J, Hatsopoulos NG (2007) Congruent activity during action and action observation in motor cortex. J Neurosci 27:13241–13250

Umiltà MA, Kohler E, Gallese V, Fogassi L, Fadiga L, Keysers C, Rizzolatti G (2001) "I know what you are doing": a neurophysiological study. Neuron 32:91–101

Umiltà MA, Escola L, Intskirveli I, Grammont F, Rochat M, Caruana F, Jezzini A, Gallese V, Rizzolatti G (2008) When pliers become fingers in the monkey motor system. Proc Natl Acad Sci U S A 105:2209–2213

Visalberghi E, Fragaszy DM (1990) Do monkeys ape? In: Parker ST, Gibson KR (eds) "Language" and intelligence in monkeys and apes. Cambridge University Press, Cambridge, pp 247–273

Williams JHG, Whiten A, Suddendorf T, Perrett DI (2001) Imitation, mirror neurons, and autism. Neurosci Biobehav Rev 25:287–295

Woolsey CN, Settlage PH, Meyer DR, Sencer W, Pinto Hamuy T, Travis AM (1952) Patterns of localization in precentral and "supplementary" motor areas and their relation to the concept of a premotor area. Res Publ Assoc Res Nerv Ment Dis 30:238–264

Chapter 6
A Molecular Basis for Intrinsic Muscle Properties: Implications for Motor Control

Kiisa C. Nishikawa, Jenna A. Monroy, Krysta L. Powers, Leslie A. Gilmore, Theodore A. Uyeno and Stan L. Lindstedt

Contributions of Muscle to Motor Control

Muscles serve a variety of functions during movement, not only shortening to provide actuation but also stabilizing joints, storing and recovering elastic potential energy, and even absorbing energy (Full and Koditschek 1999; Dickinson et al. 2000; Roberts and Azizi 2011). Over the past 20 years, the idea that muscles not only produce movement but also contribute to control of movement has become well established (Chiel and Beer 1997; Loeb et al. 1999; Nichols et al. 1999; Wagner and Blickhan 1999). Motor control thus comprises not only descending input from the nervous system and proprioceptive feedback, but also muscle viscoelastic properties, body dynamics and interactions with the environment (Hogan 1985; Chiel and Beer 1997; Wagner and Blickhan 1999; Monroy et al. 2007).

Dynamic regulation of muscle stiffness during perturbations is a long known function of proprioceptive sense organs (i.e., muscle spindles and Golgi tendon organs) and spinal reflexes (Matthews 1959). If muscles could also regulate stiffness dynamically, then they would play an important role in motor control. In fact, the nonlinear, viscoelastic behavior of muscles provides instantaneous dynamic tuning of stiffness during load perturbations (Slager et al. 1998). In classic experiments on soleus muscles of decerebrate cats, Nichols and Houk (1976) demonstrated that both sensory reflexes and muscle intrinsic properties regulate muscle stiffness in response to load perturbations. They found that denervated muscles respond instantaneously to perturbations, becoming stiffer during stretch and more compliant during unloading. After a delay of ~20 in cat soleus, the slower acting reflexes blend seamlessly with intrinsic muscle properties by adjusting muscle firing rates and recruiting additional motor units to match the altered load (Matthews 1959). These classic experiments

K. C. Nishikawa (✉) · J. A. Monroy · K. L. Powers · L. A. Gilmore · S. L. Lindstedt
Department of Biological Sciences, Northern Arizona University,
Flagstaff, AZ 86011-5640, USA
e-mail: Kiisa.Nishikawa@nau.edu

T. A. Uyeno
Department of Biology, Valdosta State University, Valdosta, GA 31698-0015, USA

M. J. Richardson et al. (eds.), *Progress in Motor Control*,
Advances in Experimental Medicine and Biology 782,
DOI 10.1007/978-1-4614-5465-6_6,

thus demonstrated that the intrinsic viscoelastic properties of muscle are critically important in stabilizing perturbed movements during the ~20 ms prior to the arrival of sensory feedback, and also at the limits of muscle recruitment when muscle force is near its minimum or maximum values and reflexes are least effective at modulating force output (Nichols and Houk 1976). The importance of muscle's instantaneous contributions to motor control is vividly illustrated by imagining an antelope attempting to outrun a lioness, when the pace is fast and any misstep, however small, is fatal.

Since this pioneering work, numerous examples have demonstrated a role for muscle intrinsic properties in stabilizing movement. In spinal frogs, perturbations applied during hindlimb wiping movements are compensated, so that the limb reaches the target in spite of the perturbation. In both intact and deafferented frogs, the hindlimb path after perturbation converges with the unperturbed path, such that the final position is always the same (Richardson et al. 2005). When guinea fowl run over rough terrain, they maintain stability by changing their posture to control velocity. Rapid changes in posture are due to muscle intrinsic properties. This simple mechanism allows for guinea fowl to absorb energy and slow down in response to a drop in terrain (Daley and Biewener 2006; Daley et al. 2009). These results suggest that compensation for perturbations is accomplished by muscle intrinsic properties.

During feeding in frogs, the mouth-opening muscles are pre-loaded prior to movement. During ballistic prey capture, recovery of elastic energy from the muscles and tendons, stored during pre-loading, determines the amplitude and speed of mouth opening (Lappin et al. 2006). These results suggest that intrinsic muscle properties not only provide stability during perturbations, but also determine the amplitude and velocity of ballistic movements.

The nonlinear, intrinsic viscoelastic properties of active muscle are best illustrated in isolated muscles as they are stretched and shortened at constant velocity (e.g., isovelocity experiments, Sandercock and Heckman 1997; Fig. 6.1). During constant velocity stretch, muscle force increases faster in the first 20 ms than during the next 50 ms of the stretch. Likewise, muscle force decreases faster initially during shortening (Fig. 6.1). Rack and Westbury (1974) were among the first to describe this time- and velocity-dependent viscoelastic behavior of muscles, in which stiffness is high initially, followed by yielding. As there were, at the time, no other candidates to whom this behavior could be attributed, they viewed it as a property of the cross-bridges and termed it the *short-range stiffness*.

In addition to this rapid response, there are also longer-lasting changes in the force output of a muscle following stretch or shortening. After stretch, muscles exhibit "force enhancement", an increase in force that persists after stretching has stopped. Likewise, "force depression" is a decrease in force that persists after shortening has stopped (Fig. 6.1). These isovelocity experiments and others like them demonstrate that the force output of muscle depends not only on the activation history of a muscle, but also its movement history and ongoing interactions with the environment. Due to the history dependence of force output, the traditional isometric length–tension and force–velocity relationships are insufficient to predict muscle force output during actual movements (Sandercock and Heckman 1997; Nichols and Cope 2004).

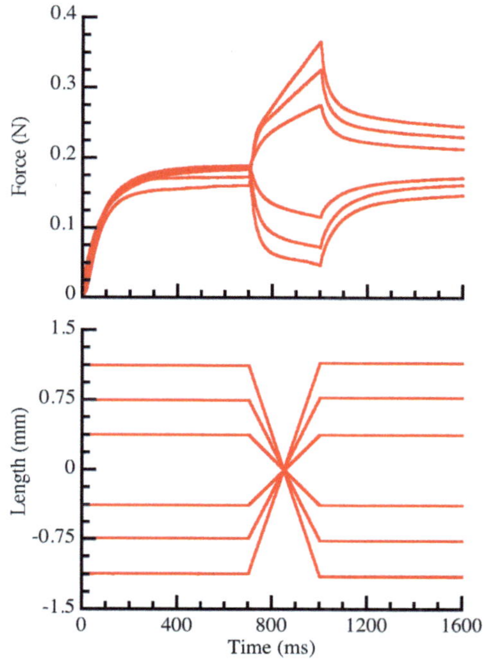

Fig. 6.1 Force (*above*) and length (*below*) data recorded during an isovelocity experiment on a single mouse soleus muscle. The muscle was first stimulated isometrically for 700 ms then stretched or shortened for 300 ms. Traces illustrate the nonlinear, time-dependent and history-dependent viscoelastic behavior of the active muscle

Not only extrafusal muscle fibers, but also the intrafusal fibers of the muscle spindle apparatus exhibit nonlinear, viscoelastic and history-dependent behavior and thus contribute to motor control (Nichols et al. 1999; Huyghues-Despointes et al. 2003a, b; Haftel et al. 2004). Whereas history-dependent behavior affects force output of extrafusal fibers, it appears that the reflex gain of spindle afferents is graded by the amplitude of prior movements in intrafusal fibers (Nichols et al. 1999).

The ability of muscles to adjust their stiffness to changes in load is important for several reasons. First, loads are imposed on a muscle by its environment, not only including reaction forces that result from interactions with external objects, but also loading imposed by the activation of antagonistic muscles as well as inertial and even coriolis forces from the musculoskeletal system. The muscles manage interactions with the environment by virtue of their nonlinear viscoelastic properties.

The fact that a mathematical representation of these interaction forces is complex (Hogan 1985) suggests that the responses of muscles to changing loads may be learned, rather than computed, and in fact in the fastest moving robots, the tuning of feedforward control to emergent body dynamics can sometimes be accomplished only by trial and error (Koditschek et al. 2004).

Despite recognition of the importance of muscle intrinsic properties to motor control, a theoretical framework that accounts for these muscle properties remains largely undeveloped. The widely accepted theory of muscle contraction, the "sliding-filament–swinging cross-bridge" theory, explains muscle contraction as resulting from the interaction between two motor proteins, myosin and actin, which are arrayed in thick and thin filaments within muscle sarcomeres (Fig. 6.2). Briefly, in this theory, overlap between the sliding filaments determines the active muscle force (Gordon et al. 1966). When a muscle is activated, myosin cross-bridges bind to actin, hydrolyze ATP, and undergo a deformation (swinging) that translates the thin filaments (Huxley 2004), producing muscle force.

However, the sliding-filament–swinging cross-bridge theory and the muscle models derived from it (i.e., Hill-Zajac, length–tension and force–velocity based models; commonly used in muscle simulations) fail to account for history dependent behavior (Sandercock and Heckman 1997; Herzog et al. 2008). Despite decades of intensive research, the molecular basis for these intrinsic properties of muscle has eluded explanation since their original observation in the early 1950s (Abbott and Aubert 1952; Herzog et al. 2008). In the absence of a plausible mechanism, phenomenological models have been used to describe the nonlinear viscoelastic behavior of muscle (Forcinito et al. 1998; Cheng et al. 2000; Lin and Crago 2002). However, these are poor substitutes for a deeper understanding of the underlying mechanisms.

We recently proposed a novel molecular mechanism, the "winding filament" hypothesis that accounts for the viscoelastic properties of active muscle (Nishikawa et al. 2011). Here, we explore the implications of the winding filament hypothesis for informing our understanding of the contributions of muscle intrinsic properties to motor control. We first review the structure and function of titin within muscle sarcomeres. Next, we describe the details of the winding filament hypothesis. Finally, we end by discussing the implications of this hypothesis for understanding the muscle's contributions to motor control.

Titin Structure and Function

The largest known protein, titin (also known as connectin), was also one of the last muscle proteins to be discovered (Maruyama et al. 1976), despite the fact that it is the third-most abundant protein in striated muscle. Although the existence of titin-like fibers was inferred in early structural studies (Huxley and Hanson 1954), titin was discovered more than 20 years after development of the sliding filament theory (Maruyama et al. 1976). For this reason, the development of the sliding-filament–swinging cross-bridge theory proceeded without considering titin.

Titin spans an entire half-sarcomere (~1 mm) from Z-disk to M-line (Gregorio et al. 1999). The overlap of titin molecules in both Z-disks and M-lines produces a titin filament system that is continuous among the entire length of a muscle fiber. Early studies of titin established its roles in maintaining sarcomere integrity and contributing to passive tension (Linke et al. 1998). Current work focuses on titin's roles in regulating myofibrillar assembly (Gregorio et al. 1999) and cell signaling (e.g., Krüger and Linke 2011).

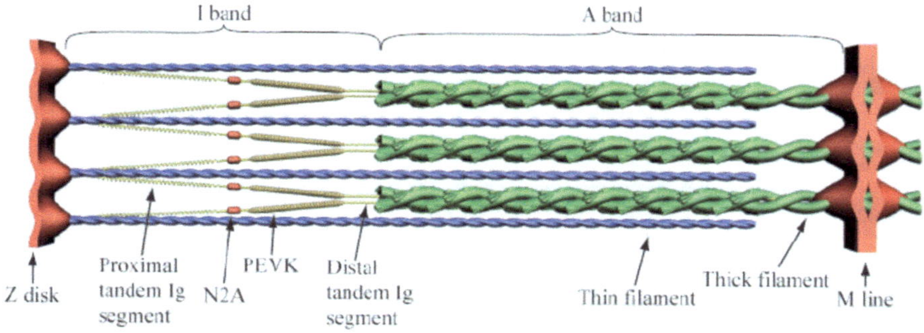

Fig. 6.2 Schematic diagram of a skeletal muscle half-sarcomere, illustrating the layout of titin (*yellow* with *red* N2A segment). Each titin molecule is bound to the thin filaments (*blue*) in the I-band, and to the thick filaments (*green*) in the A-band. For simplicity, thick filaments are illustrated as double-stranded, whereas in vertebrate skeletal muscle, they appear to be triple-stranded. The N2A region is located between the proximal tandem Ig segment and the PEVK segment. (Reprinted from Nishikawa et al. 2011)

Titin's Role in Muscle Passive Tension

The I-band region of titin (Fig. 6.2) is elastic and extends when the sarcomere is stretched, giving rise to passive muscle force (Labeit et al. 2003; Linke et al. 1998). In skeletal muscle, the I-band region of titin is composed of two serially linked spring elements: tandem immunoglobulin (Ig) domains and the PEVK segment (named for its most common amino acids). At relatively short sarcomere lengths, passive stretch straightens the folded tandem Ig domains with little change in passive tension. At longer sarcomere lengths, the PEVK segment elongates and passive tension increases steeply. Within the physiological range of sarcomere lengths, elongation of the PEVK segment largely determines the passive elasticity of skeletal muscle fibers (Linke et al. 1998).

Is There a Role for Titin in Active Muscle?

It has frequently been suggested that titin could function as a spring not only in resting muscles but also in active muscles (Bagni et al. 2002, 2004; Labeit et al. 2003; Reich et al. 2000). As yet, no compelling mechanism has been offered for how titin could play such a role. In resting muscle, titin is far too compliant to contribute significantly to active muscle force (Campbell and Moss 2002). However, several studies have demonstrated that titin stiffness increases in the presence of Ca^{2+}. In active muscle fibers, Ca^{2+} influx increases the tension and stiffness of a non-cross-bridge structure, possibly titin (Bagni et al. 2002, 2004). Ca^{2+} influx increases the stiffness of PEVK fragments as well as muscle fibers (Labeit et al. 2003). Nevertheless, the effects of Ca^{2+} on titin stiffness observed in these studies are $\sim$10 times too small to account for the observed increase in stiffness of muscle fibers upon calcium activation.

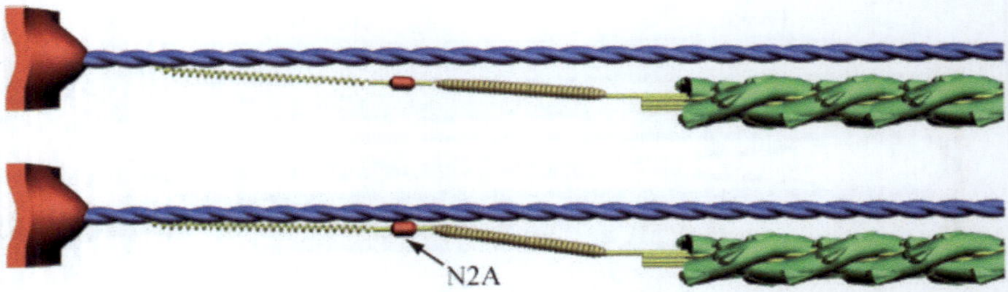

Fig. 6.3 Schematic diagram illustrating the hypothesis that titin is engaged mechanically with Ca^{2+} influx upon muscle activation. (*Above*) resting sarcomere at slack length at low Ca^{2+} concentration (pCa = 9). Titin binds to the thin filaments only near the Z-disk. (*Below*) Upon Ca^{2+} influx (pCa = 4.5), N2A binds to the thin filaments (*blue*) in the I-band, which shortens and stiffens the titin spring in active sarcomeres. (Reprinted from Nishikawa et al. 2011)

Titin has also been implicated in the increase of passive force following deactivation of actively stretched muscle fibers. In myofibrils in which active force production was prevented by removal of troponin C, a Ca^{2+} induced increase in titin-based stiffness was observed, but the increase was also too small to account for passive force enhancement (Joumaa et al. 2008). The results suggest that passive force enhancement requires not only Ca^{2+} influx, but also active force production.

In an innovative series of experiments, Leonard and Herzog (2010) stretched myofibrils, both passive and active, far beyond overlap (i.e., sarcomere lengths up to 6 μm) of the thick and thin filaments (Leonard and Herzog 2010). In these experiments, they found evidence for both an activation-dependent and a force-dependent increase in titin stiffness. At the longest lengths, the difference in stiffness between active vs. passive myofibrils was substantial. Taken together, *these experiments demonstrate that, in active muscle, titin stiffness is increased by Ca^{2+} influx and force development.*

The Winding Filament Hypothesis

Our recent "winding filament" hypothesis (Nishikawa et al. 2011) proposes that the giant, elastic titin protein is first engaged mechanically during Ca^{2+} activation in skeletal muscle, and the cross-bridges then wind titin on the thin filaments, storing elastic potential energy during force development. Storage and recovery of elastic energy in titin accounts for the time- and history-dependent behavior of active muscles.

Mechanical Engagement of Titin Upon Ca^{2+} Activation

Titin is a huge, multidomain protein that corresponds roughly in size to a thousand average-sized protein. Within this giant protein, the N2A region of titin (Fig. 6.3) is in an ideal position to modulate titin stiffness through Ca^{2+} dependent binding to

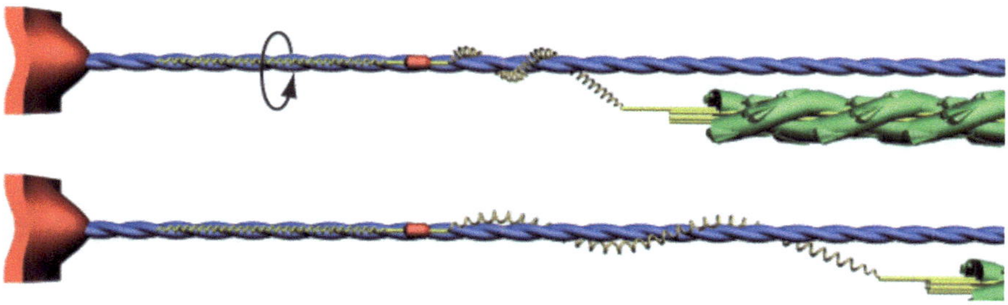

Fig. 6.4 Schematic diagram illustrating how cross-bridge cycling results in titin winding. (*Above*) Cycling of the cross-bridges winds PEVK on the thin filaments (*arrow* indicates direction of rotation). The winding angle depends only on sarcomere geometry. (*Below*) Stretch of an active sarcomere extends the PEVK segment and enhances the active force. (Reprinted from Nishikawa et al. 2011)

thin filaments. Binding of titin to actin at this location would eliminate low-force straightening of proximal tandem Ig domains in the I-band that normally occurs upon passive stretch of myofibrils at slack length (Linke et al. 1998). Furthermore, when Ca^{2+} activated sarcomeres are stretched, the PEVK segment of titin (Fig. 6.3) will elongate at high force. If Ca^{2+} dependent binding between N2A titin and thin filaments could be prevented, then active force production should decrease at short sarcomere lengths because any strain that developed in titin would straighten the tandem Ig segments at low force rather than extend the PEVK segment at higher force. Thus, the contribution of titin to the total active force would be reduced.

Thin Filament Rotation and Titin Winding

In active muscle sarcomeres, cross-bridges likely rotate as well as translate the thin filaments (Nishikawa et al. 2011; Fig. 6.4). Given the structure of the thick and thin filaments, maintenance of stereo specific binding between an actin monomer and its three neighboring thick filaments requires the thin filaments to rotate as the myosin heads translate the thin filaments toward the M-line (Morgan 1977).

As titin is bound to thick filaments in the A-band and to thin filaments in the Z-disk (Funatsu et al. 1993), rotation of thin filaments by the cross-bridges must inevitably lead to winding of titin upon them. Rotation of the thin filaments by the cross-bridges would also produce a torque in alpha-actinin in the Z-disk. Winding of titin on the thin filaments is predicted to change the length and stiffness of PEVK, storing elastic potential energy during isometric force development and active stretch. This energy could be recovered during active shortening.

Unwinding of titin from the thin filaments could be prevented by electrostatic interactions between titin's PEVK segment and the thin filaments (Bianco et al. 2007). Spontaneous dissociation rates of PEVK bound to actin are low, and the force required to break the bonds is approximately equal to the force required to break an actomyosin cross-bridge. Unwinding of PEVK from the thin filaments is

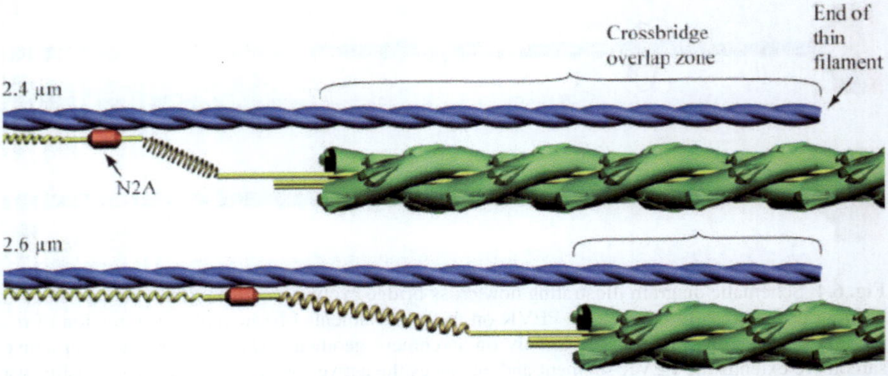

Fig. 6.5 Schematic diagram illustrating the contribution of titin to the force–length relationship. Imagine a muscle or muscle fiber that is stretched passively, and then activated at different lengths. Upon calcium influx, N2A titin (*red*) will bind to the nearest actin monomer in the thin filament (*blue*). Once N2A binds, the active elastic properties will be determined by PEVK titin and will be invariant across a range of lengths until a length is reached at which PEVK titin is extended passively before activation. As long as the binding site for N2A titin depends only on the sarcomere length at the time of activation, then a plateau is predicted in active force. For example, in rabbit psoas muscle a plateau is predicted at sarcomere lengths between 2.4 μm (*above*) and 2.6 μm (*below*). (Reprinted from Nishikawa et al. 2011)

hypothesized to occur during active shortening at low loads when the combined PEVK-actin and cross-bridge forces are too low to hold the torques in titin and alpha-actinin, as well as during muscle relaxation.

Implications For Understanding Motor Control

Here, we address implications of the winding filament hypothesis for understanding motor control. First, we discuss how mechanical engagement of the titin spring upon Ca^{2+} activation provides a mechanism by which nearly invariant contractile and viscoelastic properties can be produced regardless of the initial sarcomere length at which the muscles are activated. Next, we discuss how winding of titin on the thin filaments upon activation changes a muscle's equilibrium position and stiffness as a function of muscle recruitment. These changes, in turn, produce forces that move the limbs to their final position regardless of unexpected perturbations.

Length Invariance of Muscle Contractile and Elastic Properties

The idea that titin is engaged mechanically when N2A binds to the thin filaments upon Ca^{2+} activation has several important implications for understanding the contribution of muscle to motor control. If N2A titin can bind to a thin filament at multiple locations along its length (Fig. 6.5), then muscle contractile (e.g., force, velocity) and viscoelastic properties will remain relatively constant despite increases in sarcomere length (Edman 1979). The relative constancy of these properties with muscle length has important implications for control of movement. For example, Asatryan and Feldman (1965) demonstrated that, during involuntary arm movements elicited by unloading, as well as voluntary arm movements produced intentionally, the final position of the human arm is controlled by varying the position at which the muscles are activated. Once activated, the nonlinear viscoelastic properties of the muscles move the arm to the final position. The relative constancy of muscle viscoelastic properties across a range of muscle lengths ensures that the passive dynamics are predictable, as well as independent of the joint angle (Feldman and Levin 2009).

Motors vs. Springs: Time- and History-Dependent Properties of Active Muscle

The history-dependent properties of active extrafusal and intrafusal muscle fibers are exactly those expected of nonlinear, time-dependent springs, which produce greater tensile force when stretched and less tensile force when shortened, in proportion to the change and rate of change in length. However, within the framework of the sliding-filament theory, muscles are viewed primarily as motors. Hence, few of the ideas that have been proposed to explain the history-dependent effects deal explicitly with spring properties (see e.g., Rassier and Herzog 2004). Mechanisms of force enhancement during active stretch as well as mechanisms of force depression during shortening have invoked processes that affect the internal work done by the myosin heads during cross bridge cycling (Herzog 1998; Nichols and Cope 2004). These ideas share the common theme that the proposed mechanism interferes with the ability of the cross-bridges to produce force.

In the winding filament hypothesis, both the time dependence and history-dependence of muscle force are viewed as viscoelastic properties associated with the titin spring in muscle sarcomeres. During active stretch, muscle force increases rapidly to values up to nearly twice the maximum isometric force. The force then decays rapidly to a steady state value that increases with the amplitude of the stretch and with sarcomere length. In the winding filament hypothesis, the work done in stretching a muscle will extend titin, storing elastic strain energy. This added force increases with the distance stretched (Nishikawa et al. 2011).

During active shortening, muscle force decreases rapidly and then returns more slowly to a steady state level that depends upon both the amplitude and velocity of shortening. In the winding filament hypothesis, energy stored in titin during isometric force development will be converted to kinetic energy during shortening, and the muscle force will decrease in direct proportion to the distance shortened. The velocity dependence of force depression results from the velocity-dependent unwinding of titin from the thin filaments (Nishikawa et al. 2011).

To demonstrate how the winding filament model accounts for history dependent properties of active muscle, we developed a kinematic model (Fig. 6.6) to quantify the effects of thin filament rotation on titin during isometric force development and

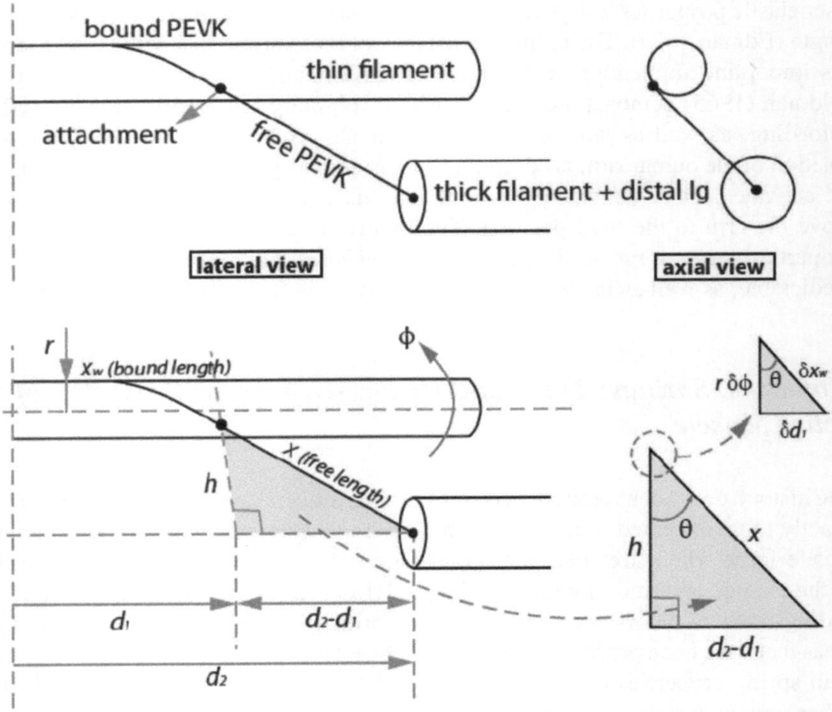

Fig. 6.6 Kinematics of titin winding. Winding angle (θ) is the angle formed between the titin filament and a line (h) parallel to the Z-disk. In the model, the winding angle is determined by sarcomere geometry and increases with sarcomere length. As the winding angle (θ) increases, the length of free titin (x) will decrease for a given angle of thin filament rotation (ϕ). d_1 distance from Z-disk to the point at which bound PEVK becomes free, d_2 distance from Z-disk to distal (C-terminal) end of PEVK, r radius. (Reprinted from Nishikawa et al. [2011])

active stretch. The model is based on a sarcomere structure similar to rabbit psoas muscle (Nishikawa et al. 2011). The model assumes that winding of titin on the thin filaments proceeds until the radial component of the cross-bridge force is equal to the sum of the radial forces in titin and alpha-actinin. As the force develops, the length of bound titin that is wound upon the thin filaments increases, increasing strain and stiffness in the free portion of titin (Fig. 6.6). When active sarcomeres are lengthened by the application of an external force, the work done in elongating free titin is stored as elastic potential energy, resulting in force enhancement at low energy cost.

Increasing strain and stiffness of titin due to thin filament rotation depends on the winding angle of titin upon the thin filament (Fig. 6.6). The winding angle (θ) is defined as the angle formed between the titin filament and a line (h) parallel to the Z-disk. In the model, the winding angle is determined by sarcomere geometry, and increases with sarcomere length. As the thin filament rotation angle (ϕ) increases, the length of the free titin segment decreases and the stress in this segment increases, thereby increasing its effective stiffness. The edge between free and bound titin will also advance toward the m-line, reducing the titin strain.

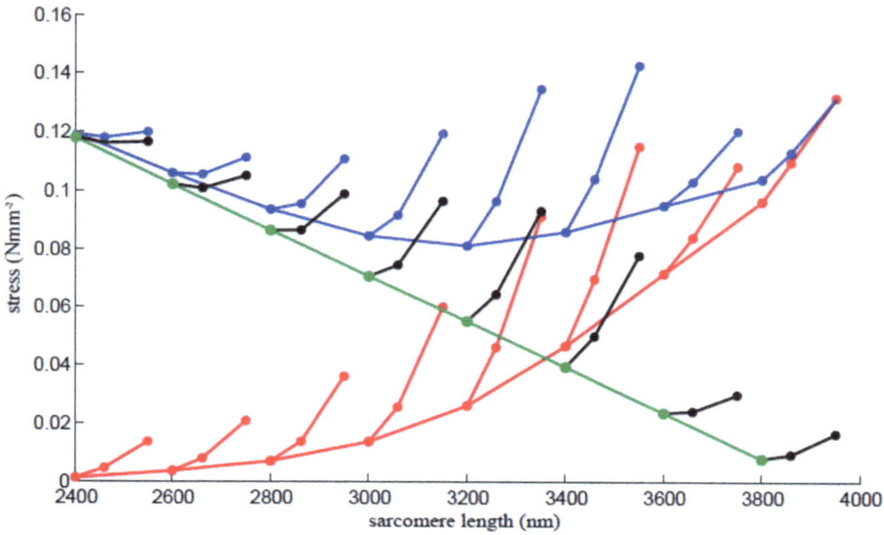

Fig. 6.7 Simulation of residual force enhancement on the descending limb of the force-length relationship. Predicted axial stress due to cross-bridges (*green*) and titin (*red*). Total axial stress (*blue*) is the sum of axial stress due to cross-bridges and titin. Baselines show steady state isometric stress. Branches show increased stress due to stretch. Residual force enhancement (*black*) is the increase in force due to active stretching above the isometric force at the corresponding length. (Reprinted from Nishikawa et al. 2011)

A nonlinear ordinary differential equation (ODE) was used to simulate the kinematics of titin winding and the resulting axial forces for a given profile of thin filament rotation $\phi(t)$ and sarcomere geometry. In the axial direction, the total force is the sum of the axial forces produced by titin and the cross-bridges. In the axial plane, the sum of the torques due to radial forces produced by titin in the I-band and alpha-actinin in the Z-disk are equal and opposite to the torque produced by the cross-bridges (Nishikawa et al. 2011).

Using this model, we simulated the force enhancement on the descending limb of the force–length relationship by alculating the axial forces produced by the cross-bridges and titin in sarcomeres activated at different initial lengths, and then stretched while active (Fig. 6.7). The results are qualitatively similar to experimental observations (Edman et al. 1982). These results demonstrate that the winding filament hypothesis accounts for the observed pattern of force enhancement in actively stretched muscles.

Motor Control and Higher Brain Centers

Theories of motor control abound and no clear consensus has emerged (Ajemian and Hogan 2010). Some workers adopt a hierarchical view of motor control (Cheng et al. 2000), in which higher brain centers (e.g., motor cortex) encode intended movements at a more abstract level (e.g., intended movement direction) and in a retinocentric coordinate frame (Georgopoulos 1986). At lower levels in the hierarchy (e.g., spinal cord), intended movements are encoded at more concrete levels (e.g., joint torque) and reference frames that are increasingly closer to the muscles that actuate the movements (see e.g., Flanders et al. 1992). Other workers have noted that feedforward control is actually simplified when the nonlinear properties of multijoint systems and intrinsic viscoelastic properties of muscle are taken into account (Hogan 1985; Todorov 2000).

A common theme of all current theories of motor control is that the feedforward controller must anticipate the nonlinear viscoelastic properties of the actuators in order to produce an intended movement. In fact, several recent neurophysiological studies suggest that the human brain anticipates the nonlinear viscoelastic properties of its muscle actuators in the neurally encoded control signals that produce voluntary movements (Feldman and Levin 2009).

The equilibrium point hypothesis (Feldman and Levin 2009) is a case in point. Asatryan and Feldman (1965) demonstrated that, the final position of the human arm during involuntary arm movements elicited by unloading and voluntary arm movements produced intentionally, is controlled by varying the initial position at which the muscles are activated. Once activated, the nonlinear viscoelastic properties of the muscles interact with length feedback to move the arm to the final position. Using transcranial magnetic stimulation to measure motor-evoked potentials, Raptis et al. (2010) and Sangani et al. (2011) showed that the human motor cortex participates in specifying the initial arm position at which the muscles are activated.

The winding filament hypothesis provides realistic biological mechanisms for implementing this simple control strategy. The engagement of the titin spring upon muscle activation provides a mechanism by which nearly invariant muscle force output can be produced when the muscles are activated at varying initial positions. The winding of titin on the thin filaments upon activation provides for changes in a muscle's characteristic length and stiffness as a function of muscle recruitment, which in turn provides the forces that move the limbs to their final positions regardless of unexpected perturbations.

Conclusion

The sliding-filament–swinging cross-bridge theory views muscles primarily as motors. Traditional hill-zajac-type muscle models based on this theory emphasize the length–tension and force–velocity properties of muscle. These models fail to predict movement dynamics because they ignore the history dependence of force output. In contrast, muscle fibers, both extrafusal and intrafusal, actually behave as nonlinear, self-stabilizing controllers that become stiffer when the external load increases and more compliant when the load decreases (Lappin et al. 2006; Monroy et al. 2007). When the load changes unexpectedly, muscle stiffness adjusts instantly without requiring neural input (Nichols and Houk 1976). In our winding filament hypothesis, the nonlinear viscoelastic properties of muscle are due to (1) Ca^{2+} activation of titin, which mechanically engages the titin spring; and (2) cross-bridge winding of titin on the thin filaments, which stores elastic energy in titin and provides viscoelastic forces that set the equilibrium position of the mechanical system.

During perturbations, intrinsic muscle properties provide stability by adjusting their stiffness instantaneously to changes in load. Thus, the muscles themselves are largely responsible for controlling the interaction between the body and the environment, as well as managing interactions between antagonistic muscles that interact via their loads. During planned movements, these intrinsic properties must be anticipated by the central nervous system, so that descending commands result in the intended movements.

It seems doubtful that a cohesive theory of motor control can be developed in the absence of a predictive model of muscle dynamics, since the central nervous system must necessarily take these into account in planning and anticipating movement. Thus, we believe that the winding filament hypothesis can fill existing gaps in our understanding of motor control. Furthermore, by providing a biological mechanism for muscle-intrinsic viscoelastic properties, the winding filament hypothesis holds great promise for inspiring the design of a new generation of actuators and prostheses that, like muscles, will exhibit self-stabilization based on variable, nonlinear compliance.

Acknowledgments We thank Michael Richardson, Michael Riley, and Kevin Shockley for inviting us to participate in the symposium and for editing this book. T. Richard Nichols and Cinnamon Pace provided helpful comments on the manuscript. Our research was supported by grants IOS-0732949, IIS-0827688, and IOS-1025806 from the National Science Foundation, and by TRIF Growing Biotechnology grants from Northern Arizona University.

References

Abbott BC, Aubert XM (1952) The force exerted by active striated muscle during and after change of length. J Physiol 117(1):77–86

Ajemian R, Hogan N (2010) Experimenting with theoretical motor neuroscience. J Motor Behav 42:333–342

Asatryan DG, Feldman AG (1965) Functional tuning of nervous system with control of movement or maintenance of a steady posture—I. Mechanographic analysis of the work of the joint on execution of a postural task. Biofizika 10:837–846

Bagni MA, Cecchi G, Colombini B, Colomo F (2002) A non-cross-bridge stiffness in activated frog muscle fibers. Biophys J 82:3118–3127

Bagni MA, Colombini B, Geiger P, Berlinguer Palmini R, Cecchi G (2004) Non-crossbridge calcium-dependent stiffness in frog muscle fibers. Am J Physiol Cell Physiol 286:C1353–1357

Bianco P, Nagy A, Kengyel A, Szatmari D, Martonfalvi Z, Huber T, Kellermayer MS (2007) Interaction forces between F-actin and titin PEVK domain measured with optical tweezers. Biophys J 93:2102–2109

Campbell KS, Moss RL (2002) History-dependent mechanical properties of permeabilized rat soleus muscle fibers. Biophys J 82:929–943

Cheng EJ, Brown IE, Loeb GE (2000) Virtual muscle: a computational approach to understanding the effects of muscle properties on motor control. J Neurosci Meth 101:117–130

Chiel HJ, Beer RD (1997) The brain has a body: adaptive behavior emerges from interactions of nervous system, body and environment. TINS 20:553–557

Daley MA, Biewener AA (2006) Running over rough terrain reveals limb control for intrinsic stability. Proc Nat Acad Sci 103:15681–15686

Daley MA, Voloshina A, Biewener AA (2009) The role of intrinsic muscle mechanics in the neuromuscular control of stable running in the guinea fowl. J Physiol 587:2693–2707

Dickinson MH, Farley CT, Full RJ, Koehl MAR, Kram R, Lehman S (2000) How animals move: an integrative view. Science 288:100–106

Edman KA (1979) The velocity of unloaded shortening and its relation to sarcomere length and isometric force in vertebrate muscle fibres. J Physiol 291:143–159

Edman KA, Elzinga G, Noble MI (1982) Residual force enhancement after stretch of contracting frog single muscle fibers. J Gen Physiol 80:769–784

Feldman AG, Levin MF (2009) The equilibrium-point hypothesis—past, present and future. Adv Exp Med Biol 629:699–726

Flanders M, Tillery SIH, Soechting JF (1992) Early stages in a sensorimotor transformation. Behav Brain Sci 15:309–362

Forcinito M, Epstein M, Herzog W (1998) Can a rheological muscle model predict force depression/enhancement? J Biomech 31:1093–1099

Full RJ, Koditschek DE,(1999) Templates and anchors—neuromechanical hypotheses of legged locomotion on land. J Exp Biol 202:3325–3332

Funatsu T, Kono E, Higuchi H, Kimura S, Ishiwata S, Yoshioka T, Maruyam K, Tsukita S (1993) Elastic filaments in situ in cardiac muscle: deep-etch replica analysis in combination with selective removal of actin and myosin filaments. J Cell Biol 120:711–724.

Georgopoulos AP (1986) On reaching. Ann Rev Neurosci 9:147–170

Gordon AM, Huxley AF, Julian FJ (1966) The variation in isometric tension with sarcomere length in vertebrate muscle fibers. J Physiol (Lond) 184:170–192

Gregorio CC, Granzier H, Sorimachi H, Labeit S (1999) Muscle assembly: a titanic achievement? Curr Opin Cell Biol 1:18–25

Haftel VK, Bichler EK, Nichols TR, Pinter MJ, Cope TC (2004) Movement reduces the dynamic response of muscle spindle afferents and motor neuron synaptic potentials in rat. J Neurophysiol 91:2164–2171

Herzog W (1998) History dependence of force production in skeletal muscle: a proposal for mechanisms. J Electromyogr Kinesiol 8:111–117

Herzog W, Leonard TR, Joumaa V, Mehta A (2008) Mysteries of muscle contraction. J Appl Biomech 24:1–13

Hogan N (1985) The mechanics of multi-joint posture and movement control. Biol Cybern 52: 315–331

Huxley HE (2004) Recent X-ray diffraction studies of muscle contraction and their implications. Phil Trans Roy Soc Lond B. Bio 359:1879–1882

Huxley H, Hanson J (1954) Changes in the cross-striations of muscle during contraction and stretch and their structural interpretation. Nature 173:973–976

Huyghues-Despointes CM, Cope TC, Nichols TR (2003a) Intrinsic properties and reflex compensation in reinnervated triceps surae muscles of the cat: effect of activation level. J Neurophysiol 90:1537–1546

Huyghues-Despointes CM, Cope TC, Nichols TR (2003b) Intrinsic properties and reflex compensation in reinnervated triceps surae muscles of the cat: effect movement history. J Neurophysiol 90:1547–1555

Joumaa V, Leonard TR, Herzog W (2008) Residual force enhancement in myofibrils and sarcomeres. Proc Biol Sci 275:1411–1419

Koditschek DE, Full RJ, Buehler M (2004) Mechanical aspects of legged locomotion control. Arthropod Struct Dev 33:251–272

Krüger M, Linke WA (2011) The giant protein titin: a regulatory node that integrates myocyte signaling pathways. J Biol Chem 286:9905–9912

Labeit D, Watanabe K, Witt C, Fijita H, Wu Y, Lahmers S, Funck T, Labeit S, Granzier H (2003) Calcium-dependent molecular spring elements in the giant protein titin. Proc Natl Acad Sci U S A 100:13716–13721

Lappin AK, Monroy JA, Pilarski JQ, Zepnewski ED, Pierotti DJ, Nishikawa KC (2006) Storage and recovery of elastic potential energy powers ballistic prey capture in toads. J Exp Biol 209:2535–2553

Leonard TR, Herzog W (2010) Regulation of muscle force in the absence of actin-myosin-based cross-bridge interaction. Am J Physiol Cell Physiol 299:C14–20

Lin CCK, Crago PE (2002) Neural and mechanical contributions to the stretch reflex: a model synthesis. Ann Biomed Eng 30:54–67

Linke WA, Ivemeyer M, Mundel P, Stockmeier MR, Kolmerer B (1998) Nature of PEVK-titin elasticity in skeletal muscle. Proc Natl Acad Sci U S A 95:8052–8057

Loeb GE, Brown IE, Cheng EJ (1999) A hierarchical foundation for models of sensorimotor control. Exp Brain Res 126:1–18

Maruyama K, Natori R, Nonomura Y (1976) New elastic protein from muscle. Nature 262:58–60

Matthews PBC (1959) The dependence of tension upon extension in the stretch reflex of the soleus muscle of the decerebrate cat. J Physiol 147:521–546

Monroy JA, Lappin AK, Nishikawa KC (2007) Elastic properties of active muscle—on the rebound? Exerc Sport Sci Rev 35:174–179

Morgan RS (1977) Actin rotates as myosin translates. J Theor Biol 67:769–771

Nichols TR Houk JC (1976) Improvement in linearity and regulation of stiffness that results from actions of stretch reflex. J Neurophysiol 39:119–42

Nichols TR Cope TC (2004) Cross-bridge mechanisms underlying the history-dependent properties of muscle spindles and stretch reflexes. Can J Physiol Pharmacol 82:569–576

Nichols TR, Lin DC, Huyghues-Despointes CM (1999) The role of musculoskeletal mechanics in motor coordination. Prog Brain Res 123:369–378

Nishikawa KC, Monroy JA, Uyeno TE, Yeo SH, Pai DK, Lindstedt SL (2011) Is titin a 'winding filament'? A new twist on muscle contraction. Proc Roy Soc Lond B 279:981–990

Rack PMH, Westbury DR (1974) The short-range stiffness of active mammalian muscle and its effect on mechanical properties. J Physiol 240:331–350

Raptis H, Burtet L, Forget R, Feldman AG (2010) Control of wrist position and muscle relaxation by shifting spatial frames of reference for motoneuronal recruitment: possible involvement of corticospinal pathways. J. Physiol 88:1551–1570

Rassier DE, Herzog W (2004) Considerations on the history dependence of muscle contraction. J Appl Physiol 96:419–427

Reich TE, Lindstedt SL, LaStayo PC, Pierotti DJ (2000) Is the spring quality of muscle plastic? Am J Physiol Regul Integr Comp Physiol 278:R1661–1666

Richardson AG, Slotine JE, Bizzi E, Tresch MC (2005) Intrinsic musculoskeletal properties stabilize wiping movements in the spinalized frog. J Neurosci 25:3181–3191

Roberts TJ, Azizi E (2011) Flexible mechanisms: the diverse roles of biological springs in vertebrate movement. J Exp Biol 214:353–361

Sangani SG, Raptis HA, Feldman AG (2011) Subthreshold corticospinal control of anticipatory actions in humans. Behav Brain Res 224:145–154

Sandercock TG, Heckman CJ (1997) Force from cat soleus muscle during imposed locomotor-like movements: experimental data versus Hill-type muscle predictions. J Neurophysiol 77: 1538–1552

Slager GEC, Otten E, Nagashima T, Van WIlligen JD (1998) The riddle of the large loss in bite force after fast jaw-closing movements. J Dent Res 77:1684–1693

Todorov E (2000) Direct cortical control of muscle activation in voluntary arm movements: a model. Nat Neurosci 3:391–398

Wagner H, Blickhan R (1999) Stabilizing function of skeletal muscles: an analytical investigation. J Theor Biol 199:163–179

Chapter 7
Theoretical and Methodological Issues in Serial Correlation Analysis

Didier Delignières and Vivien Marmelat

Time Series and Auto-correlation Function

Time series analysis is a relatively recent approach in experimental psychology and movement sciences. For a long time, repeated measurements of performance were performed for obtaining more accurate estimates of average performance or for assessing performance variability. From this point of view, successive performances were considered independent and their order of occurrence was not considered. In other words, successive performances were considered white noise fluctuations around a mean stable value. However, Slifkin and Newell (1998) argued that this 'white noise assumption' could be more the exception rather than the rule, and a number of experiments have evidenced the presence of correlations between successive measurements of performance. The aim of time series analyses is to reveal and characterize serial dependence in series of observations.[1]

The concept of serial dependence refers to systematic relationships between observations over time. The autocorrelation function represents a straightforward way for analyzing serial dependence. Autocorrelation is defined as the correlation of a time series with itself, considering a given lag. The auto-correlation function represents the series of auto-correlation values obtained for increasing lags. By definition, the auto-correlation function of a white noise process presents nonsignificant, close-to-zero values, revealing the absence of dependence in the series (see Fig. 7.1, *left column*).

Serial dependence can occur short term: In that case, the current value is dependent on the previous value or on a few sets of preceding values. A good example is provided by auto-regressive series, obeying the following equation:

$$y_i = \varphi y_{i-1} + \varepsilon_i \tag{7.1}$$

[1] Note that we just define a time series as a series of data ordered in time. A strict definition supposes that successive data are spaced by regular time intervals, but in most experiments the series are simply composed of ordered data.

D. Delignières (✉) · V. Marmelat
EA 2991 Movement to Health - Euromov, UFR STAPS, University Montpellier 1,
700 avenue du Pic Saint Loup, 34090 Montpellier, France
e-mail: didier.delignieres@univ-montp1.fr

M. J. Richardson et al. (eds.), *Progress in Motor Control*,
Advances in Experimental Medicine and Biology 782,
DOI 10.1007/978-1-4614-5465-6_7,

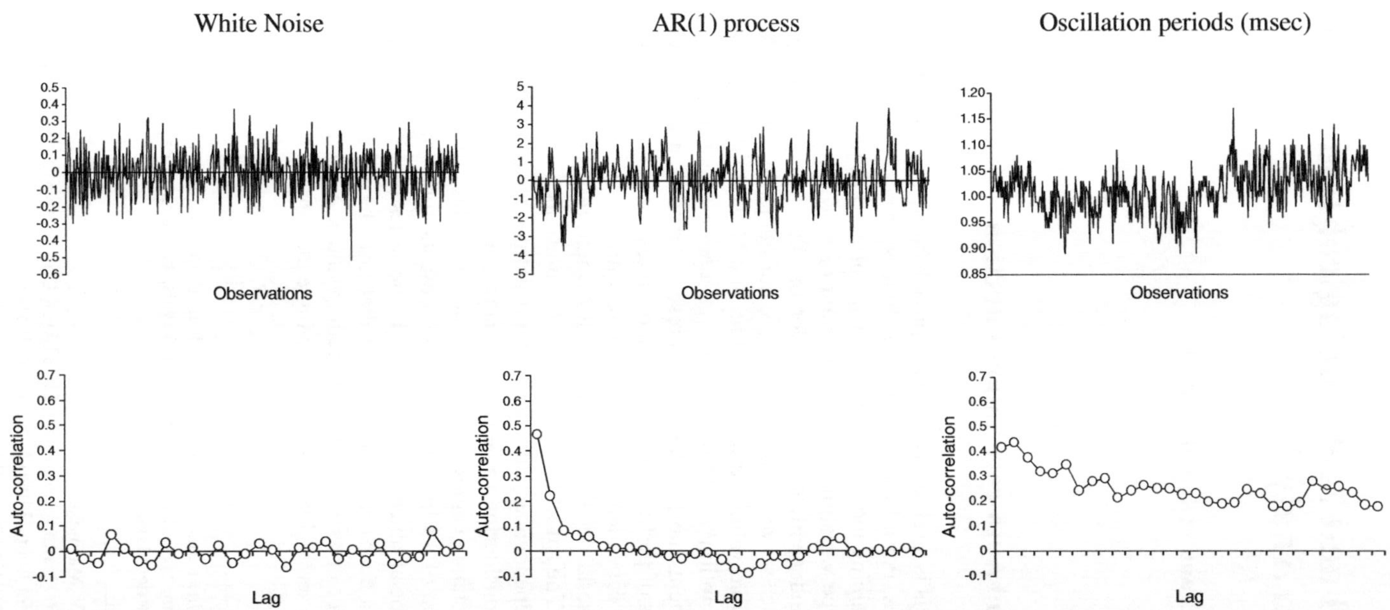

Fig. 7.1 Time series (*top row*) and auto-correlation functions (*bottom row*). *Left column*: white (uncorrelated) noise; *middle column*: one-order auto-regressive process; *right column*: experimental series of periods produced during repetitive oscillations of the forearm. (Data from Delignières et al. 2004b)

where φ is the autoregressive parameter and ε_i is a white noise process with zero mean and unit variance. In this kind of process, the current value (y_i) is defined as a fraction of the preceding value (y_{i-1}), plus a random perturbation (ε_i). The strength of the relation between successive values is entirely defined by the value of the autoregressive parameter. In this very simple process, the current value is directly dependent on the immediately preceding value. Serial correlations can appear between y_i and more distant observations (y_{i-2}, y_{i-3}, etc.), but these relations are not direct and are explained by the presence of intermediate observations (e.g., the observed relationship between y_i and y_{i-2} is explained by the effective relationships between y_i and y_{i-1}, and between y_{i-1i} and y_{i-2}).

The autocorrelation function of this kind of process typically presents a significant value at lag one (depending on the value of φ), and then a quick decay over time (see Fig. 7.1, *central column*).

Long-range Correlations

This exponential decay of the autocorrelation function, however, is rarely observed in empirical series. Often the autocorrelation function presents a very slow decay over time, suggesting that the current value presents dependencies with a large set of preceding values. For example, we present in Fig. 7.1 (*right column*) a series of periods produced during the repetitive oscillations of the forearm (Delignières et al. 2004b). The auto-correlation function presents a very slow decay over time, and remains significant over a number of lags. This persistence of auto-correlation signs the presence of long-range correlations (LRC) in the series.

Long-range correlations are characterized by the presence of dependence which tends to persist over dozens or even hundreds of data. In this kind of process, the current observation seems to keep the memory of a large set of previous observations. LRC can be understood through the fact that over multiple, interpenetrated time scales, an increasing trend in the past is likely to be followed by an increasing trend in the future, and conversely a decrease in the past is likely to be followed by a decrease in the future. This kind of process has been referred to as long-range dependence, long-term memory, fractal process, or $1/f$ noise.

Long-range correlations have been discovered in the dynamics of a number of natural and physical systems, including for example the series of discharges of the Nile River (Hurst 1951), the series of magnitudes of earthquakes (Matsuzaki 1994), the evolution of traffic in Ethernet networks (Leland et al. 1994), or the dynamics of self-esteem over time (Delignières et al. 2004a). In the domain of human movement, LRC have been evidenced in serial reaction time (Gilden 1997; Van Orden et al. 2003), in finger tapping (Gilden et al. 1995; Lemoine et al. 2006), in stride duration during walking (Hausdorff et al. 1995), or in relative phase in a bimanual coordination task (Torre et al. 2007a).

Fractional Gaussian Noise and Fractional Brownian Motion

Most often, LRC has been modeled through a family of processes introduced by Mandelbrot and van Ness (1968), namely *fractional Gaussian noises* and *fractional Brownian motions*. A formal introduction to this framework is necessary for a complete understanding of this approach. The authors first defined a family of processes they called fractional Brownian motions (fBm). The starting point of this definition is ordinary Brownian motion, a well-known stochastic process that could represent, for example, the movement of a single particle along a straight line due to a succession of random jumps. Then, an important property of Brownian motion is that its successive increments in position are uncorrelated: each displacement is independent of the former, in direction as well as in amplitude. Einstein (1905) showed that, on average, this kind of motion moves a particle from its origin by a distance that is proportional to the square root of the time.

Fractional Brownian motions (fBm) series differ from ordinary Brownian motion by the fact that in an fBm the successive increments are correlated. A positive correlation signifies that an increasing trend in the past is likely to be followed by an increasing trend in the future; in that case the series is said to be persistent. Conversely, a negative correlation signifies that an increasing trend in the past is likely to be followed by a decreasing trend, and the series is said to be antipersistent.

Mathematically, fBm is characterized by the following scaling law:

$$SD(x) \propto \Delta t^H \tag{7.2}$$

which signifies that the standard deviation of the process is a power function of the time interval (Δt) over which it was computed. H is the so-called Hurst exponent and can be any real number in the range $0 < H < 1$. Equation 7.2 describes the so-called *diffusion property* of fBm series: the higher the H, the higher the diffusion of process over time. For this point of view, ordinary Brownian motion corresponds to the special case $H = 0.5$ and constitutes the frontier between antipersistent ($H < 0.5$) and persistent fBm series ($H > 0.5$). In Fig. 7.2, we present three example fBm series for three contrasted H exponents.

Fractional Gaussian noise (fGn) represents another family, defined as the series of successive increments in an fBm. In other words, an fGn is the differentiation of an fBm and conversely the integration of an fGn gives an fBm. Each fBm series is then related to a specific fGn series, and both are characterized by the same H exponent. We present in the *bottom row* of Fig. 7.2 the fGn series corresponding to the fBm series just above. The fGn family is centered around white noise ($H = 0.5$), which represents the frontier between antipersistent ($H < 0.5$) and persistent fGn ($H > 0.5$).

These two families of processes possess fundamentally different properties: fBm series are non-stationary with time-dependent variance (diffusion property), while fGn are stationary with a constant expected mean value and constant variance over time. As previously explained, fGn and fBm can be conceived as two superimposed families, invertible in terms of differentiation and integration.

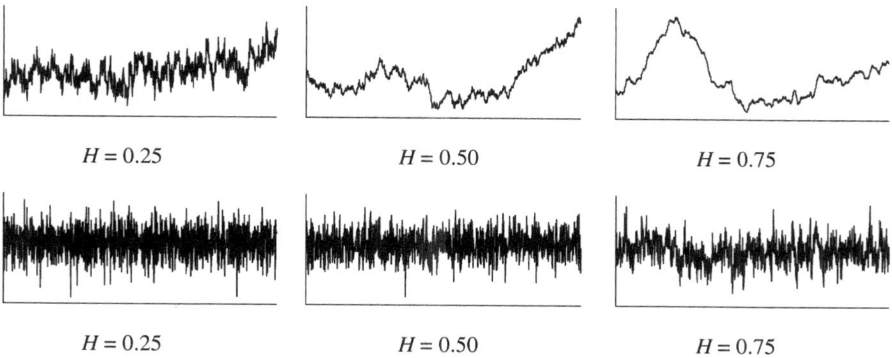

Fig. 7.2 *Top row*: Example series of fractional Brownian motions (fBm) for three typical values of the scaling exponent. The *central graph* represents an ordinary Brownian motion ($H = 0.5$). The *left graph* shows an antipersistent fBm ($H = 0.25$) and the *right graph* a persistent fBm ($H = 0.75$). The corresponding fractional Gaussian noises series (fGn) are displayed in the *bottom row*

Another useful conception is to conceive these two families as representing a continuum, ranging from the most antipersistent fGn to the most persistent fBm. This fGn/fBm continuum is characterized by the presence of scaling laws that could be expressed in the frequency or in the time domain. In the frequency domain, a scaling law relates power (i.e., squared amplitude) to frequency according to an inverse power function, with an exponent β:

$$S(f) \propto \frac{1}{f^{\beta}} \tag{7.3}$$

This scaling law is exploited by the Power Spectral Density (PSD) method that reveals β as the negative of the slope of the log–log representation of the power spectrum (Fig. 7.3). The fGn/fBm continuum is then characterized by exponents β ranging from -1 to 3 (see Fig. 7.5).

In the time domain, the typical scaling law states that the standard deviation of the integrated series is a power function of the time over which it is computed, with an exponent α. Considering a time series $x(i)$:

$$\begin{cases} y(i) = \sum_{k=0}^{i} x(k) \\ SD(y) \propto n^{\alpha} \end{cases} \tag{7.4}$$

This scaling law is exploited by the Detrended Fluctuation Analysis (DFA) that reveals α as the slope of the log–log diffusion plot (Fig. 7.4). The fGn/fBm continuum is characterized by exponents α ranging from 0 to 2 (see Fig. 7.5). Note that the scaling law expressed in Eq. 7.4 is derived from the original definition of fBm (Eq. 7.2). If the series $x(i)$ is an fGn, $y(i)$ is the corresponding fBm and α is the Hurst exponent. If $x(i)$

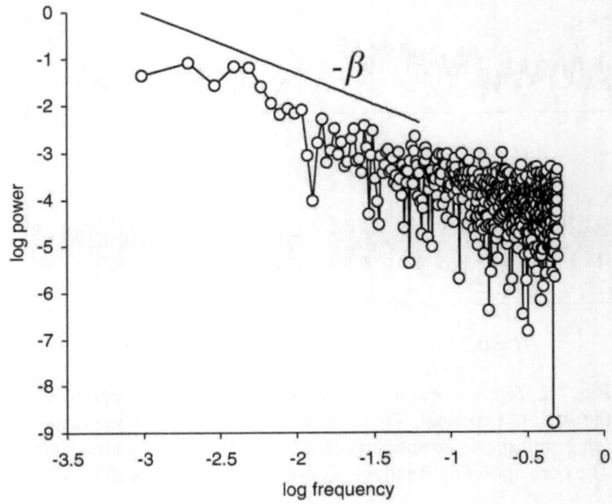

Fig. 7.3 Power spectral density analysis. The exponent β is given by the negative of the slope of the log–log representation of the power spectrum

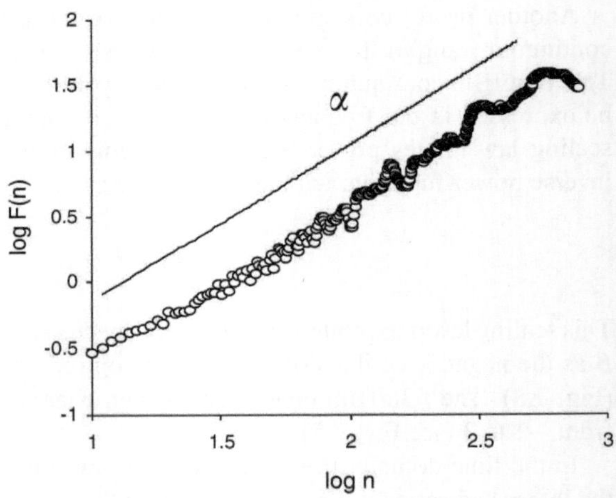

Fig. 7.4 Detrended fluctuation analysis. The exponent α is determined as the slope of the log–log diffusion plot

is an fBm, $y(i)$ belongs to another family of over-diffusive processes, characterized by exponents α ranging from 1 to 3, and in that case $\alpha = H + 1$.

Note that the different exponents characterizing these scaling laws are mutually linked by very simple equations:

For fGn series:

$$\beta = 2H - 1 \quad \text{and} \quad \alpha = H \tag{7.5}$$

For fBm series:

$$\beta = 2H + 1 \quad \text{and} \quad \alpha = H + 1 \tag{7.6}$$

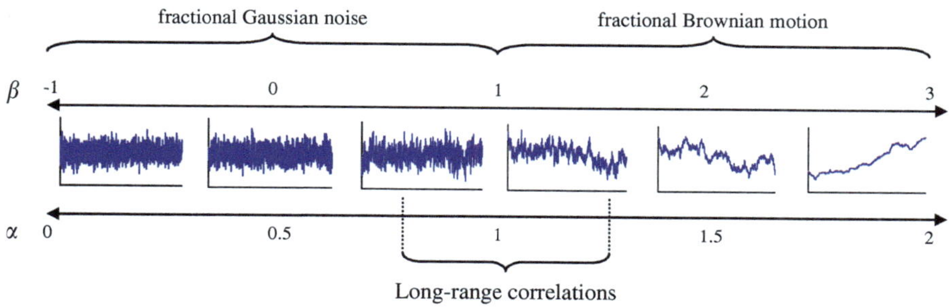

Fig. 7.5 Representation of the fGn/fBm continuum. The continuum is characterized by exponents β ranging from -1 to 3, and by exponents α ranging from 0 to 2. White noise corresponds to $\beta = 0$ and $\alpha = 0.5$, and Brownian motion to $\beta = 2$ and $\beta = 1.5$. Long range correlations are considered to appear between $\beta = 0.5$ and $\beta = 1.5$. $\beta = \alpha = 1$ corresponds to ideal $1/f$ noise

For fGn and fBm series:

$$\beta = 2\alpha + 1 \quad \text{or} \quad \alpha = \frac{\beta - 1}{2} \tag{7.7}$$

The exponents provided by PSD and DFA (β and α, respectively) are useful because they allow to unambiguously distinguish between fGn and fBm series, which could be characterized by the same H exponents. As an example, the two series presented in the *right column* of Fig. 7.2 are characterized by the same H exponent ($H = 0.25$). However, for the fGn series (*bottom row*), $\beta = -0.5$ and $\alpha = 0.25$, and for the fBm series (*top row*), $\beta = 1.5$ and $\alpha = 1.25$.

In this fGn/fBm continuum, LRC are generally considered to appear in a narrow range, between $\beta = 0.5$ and $\beta = 1.5$ (i.e., between $\alpha = 0.75$ and $\alpha = 1.25$; see Wagenmakers et al. 2004). This range is centered on $\beta = \alpha = 1$, corresponding to the ideal $1/f$ noise. Long-range correlated series present typical fluctuations, often referred to as $1/f$ fluctuations, characterized by multiple interpenetrated waves, a weak stationarity, and a marked level of roughness.

Long-range Correlations, Complexity, and Health

As previously explained, LRC have been discovered in many physical and natural systems. Science being generally based on the detection of differences (between groups or between conditions), one could question the scientific interest of studying such a ubiquitous feature. From this point of view, $1/f$ fluctuations are considered an omnipresent background noise, without discriminative power. As such, LRC could be thought just as an amazing phenomenon or a mathematics fantasy.

Another line of criticism is to consider $1/f$ fluctuations as unexpected perturbations, due to prolonged observation times. For example, the presence of LRC in tapping experiments can be supposed to reflect basically the inability of participants to keep a steady tempo, especially in long sequences. From this point of view, 1/*f* fluctuations can be thought just as undesirable as any other noise. In other words, researchers focusing on LRC just create the problem they wish to study, and 1/*f* fluctuations hide the deterministic or 'true' part of the collected signals. In sum, LRC are sometimes conceived at best a mathematical curiosity, at worst an experimental artifact (Pressing and Jolley-Rogers 1997).

A number of authors, however, have adopted a more heuristic point of view that considers 1/*f* fluctuations to reveal the structural and functional properties of the system that generated them. Considering the ubiquity of 1/*f* fluctuations in the outcome series produced by natural, physical, and biological systems, some authors have suggested that the occurrence of such fluctuation could correspond to some universal principles (Beltz and Kello 2006; Ihlen and Vereijken 2010; Kello et al. 2007, 2010). For a few years, a debate opposed these authors to the proponents of an alternative, so-called *mechanistic* approach, seeking for specific explanations and models for the presence of 1/*f* fluctuations in specific systems or situations (Torre and Wagenmakers 2009), but by now a general agreement seems to emerge favoring the *nomothetic* view considering LRC as related to universal properties (Diniz et al. 2011). From this point of view, LRC are considered to reflect the complexity of the system, defined as the flexible and adaptable coordination between its multiple components and subsystems (Diniz et al. 2011; Kello et al. 2007). The 1/*f* fluctuations do not arise from some specific component within the system but from the complex, multiplicative interactions between the multiple components acting at different time scales that compose the system. Several characteristic features of complex systems such as self-organized criticality (Van Orden et al. 2003), multiscale dynamics (Hausdorff 2005), metastability (Kello et al. 2007), or degeneracy (Delignières et al. 2011) have been advocated to play a central role in the emergence of LRC.

As previously explained, LRC seem to appear in series produced by healthy, sustainable, adaptable, and flexible systems (Bassingthwaighte et al. 1994; Goldberger et al. 2002a; Van Orden 2007; West 1990). For example, Goldberger et al. (2002a) have showed that the heart inter-beat interval series recorded in young and healthy subjects typically presented 1/*f* fluctuations. In contrast, heartbeat series appear more random in patients with a cardiac arrhythmia, and conversely more periodic and predictable in patients with severe congestive heart failure. Similar alterations of LRC have been evidenced by Hausdorff et al. (1997), who showed that stride interval series during walking presented 1/*f* fluctuations in young and healthy participants, but tended to lose their correlation structure (i.e., became more random) in elderly and in patients suffering from neurodegenerative diseases. Note that this alteration of LRC, especially in elderly, is consistent with the *loss of complexity* hypothesis proposed by Goldberger et al. (2002b).

The observation of such alterations suggests that LRC represent an optimal compromise, between order and disorder, order reflecting a too strict and rigid coordination, yielding stereotyped behavior and low adaptability, and disorder an absence of coordination. Long-range correlations can be considered a macroscopic biomarker of health, and their detection and the assessment of their alteration in specific populations or situations appear as an important scientific goal (Eke et al. 2000).

Time Series Analysis Methods

We previously evoked two of the most used methods, PSD and DFA. These methods are useful because they can be applied on both fGn and fBm series. PSD has been improved by Eke et al. (2000) with the introduction of some preprocessing operations. In a first step, the mean of the series is subtracted from each value and then a parabolic window is applied: each value in the series is multiplied by the following function:

$$W(j) = 1 - \left(\frac{2j}{N+1} - 1\right)^2 \quad \text{for } j = 1, 2, \ldots, N. \tag{7.8}$$

Thirdly a bridge detrending is performed by subtracting from the data the line connecting the first and last point of the series. The Fast Fourier Transform algorithm is applied on the resulting series. Finally the estimation of β, on the basis of Eq. 7.3, excludes the high-frequency power estimates ($f > 1/8$ of maximal frequency). This method was proven by Eke et al. (2000) to provide more reliable estimates of the spectral exponent (see also Delignières et al. 2006).

Detrended fluctuation analysis was initially proposed by Peng et al. (1993). This method exploits the scaling law expressed in Eq. 7.4, and includes a detrending procedure in order to cope with local nonstationarities in the series. The analyzed series $x(t)$ is first integrated by computing for each t the accumulated departure from the mean of the entire series:

$$X(k) = \sum_{i=1}^{k} [x(i) - \bar{x}] \tag{7.9}$$

This integrated series is then divided into nonoverlapping intervals of length n. In each interval, a least squares line is fit to the data (representing the trend in the interval). The series $X(t)$ is then locally detrended by subtracting the theoretical values $X_n(t)$ given by the regression. For a given interval length n, the characteristic size of fluctuation for this integrated and detrended series is calculated by:

$$F = \sqrt{\frac{1}{N} \sum_{k=1}^{N} [X(k) - X_n(k)]^2} \tag{7.10}$$

This computation is repeated over all possible interval lengths (in practice, the shortest length is around 10, and the largest $N/2$, giving two adjacent intervals). Typically, F increases with interval length n. A power law is expected, as

$$F \propto n^{\alpha} \tag{7.11}$$

α is expressed as the slope of a double logarithmic plot of F as a function of n.

Delignières et al. (2006) presented an evaluation of the efficiency of these methods for measuring the strength of LRC in simulated series. As can be seen in Fig. 7.6,

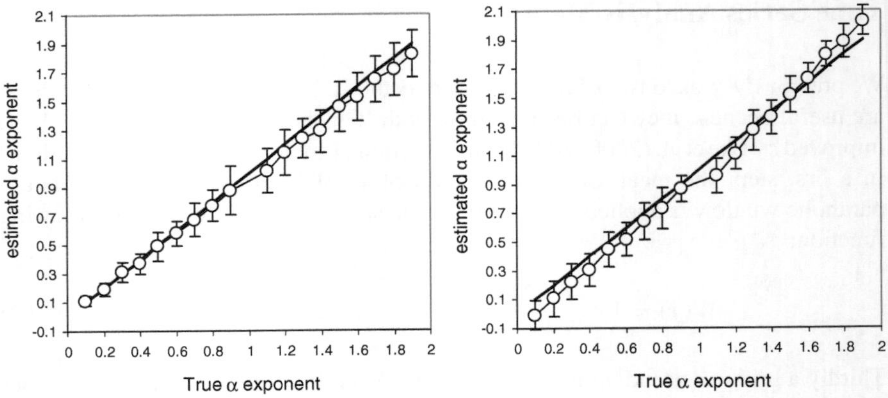

Fig. 7.6 Accuracy and variability of DFA (*left*), and PSD (*right*) estimates of scaling exponents, based on the analysis of simulated series with known exponents. The graphs represent the relationship between the exact exponents (expressed in αDFA, from $\alpha = 0.1$ to $\alpha = 0.9$ for fGn, and from $\alpha = 1.1$ to $\alpha = 1.9$ for fBm), and the mean estimates provided by the two methods. Error bars represent standard deviation. (Data from Delignières et al. 2006)

both methods gave satisfying results. DFA worked especially well with fGn signals ($0 < \alpha < 1$), but presented a systematic underestimation bias in fBm. PSD produced a slight underestimation bias in fGn series and in low diffusive fBm (especially for $\alpha = 1.1$), and conversely an overestimating bias in high diffusive fBm series ($\alpha > 1.7$). In all cases the estimates presented moderate variability levels.

Some other methods have been proposed, but they often present some limitations in their ranges of application. For example, the Rescaled Range analysis (Hurst 1965; Caccia et al. 1997) and the Dispersional analysis (Caccia et al. 1997) only work on fGn signals and gave erroneous results when applied to fBm signals. Conversely, Windowed Variance Analysis methods work only on fBm signals (Cannon et al. 1997; for a general presentation see Eke et al. 2000, 2002; Delignières et al. 2006).

It is important to note that the accuracy of the estimation of fractal exponents is directly related to the length of the analyzed series. However, Delignières et al. (2006) showed that fractal analysis methods provided acceptable results with series of 1024 or 512 points, except PSD that appeared severely affected by the shortening of series, despite the algorithmic improvements introduced by Eke et al. (2000). The results obtained with series of 256 data points were not so bad, especially with DFA. This observation is very important because of the difficulty to obtain long time series in psychological and behavioral experiments. These results suggest that a better estimate of scaling exponents could be obtained—with a similar time on the experimental task—from the average of four exponents derived from distinct 256 data point series (with an appropriate period of rest between two successive sessions) than from a single session providing 1024 data points. This could open new perspectives of research in areas that were until now reticent for using this kind of analyses.

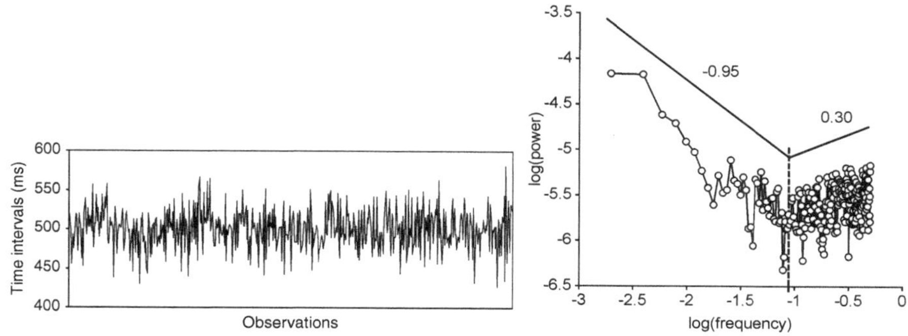

Fig. 7.7 *Right*: An example time series of inter-tap intervals collected in a self-paced finger tapping task. *Left*: Averaged log–log power spectrum. ($N = 11$, from Delignières et al. 2004b)

Short-term and Long-range Correlations

The detection of long-range correlation is not so straightforward, however, in empirical series. A number of studies have been published in the last decade, which showed that the pattern of correlation in a given signal appeared as a complex combination of short-range and long-range correlation processes.

A very classical example of such combination has been described by Gilden et al. (1995) in the series of time intervals produced during finger tapping. The application of PSD on these series produces a very specific power spectrum, with a negative slope in the low frequency region of the log–log spectrum and a positive slope in the high frequency region (Fig. 7.7).

The authors interpreted these results on the basis of the well-known model by Wing and Kristofferson (1973) suggesting that inter-tap intervals are determined by the cognitive intervals produced by an internal clock, plus a term of differenced white noise due to the motor delay that characterizes each tap.

$$I_i = C_i + M_i - M_{i-1} \tag{7.12}$$

In the initial formulation of the model, C_i and M_i were both considered as white noise sources. According to Gilden et al., the positive slope in high frequencies is due to the negative short-term correlations induced by the differenced white noise term ($M_i - M_{i-1}$). They concluded that LRC in tapping series were induced by the timekeeper (C_i), which was then supposed to possess fractal properties.

The results are more complex when tapping is performed in synchronization with a metronome. In that case the application of PSD on time interval series yields a positive slope over the entire spectrum, suggesting the presence of negative correlations in the series (Fig. 7.8).

Torre and Delignières (2008) interpreted this result on the basis of a model initially proposed by Vorberg and Wing (1996).

$$A_{i+1} = (1 - \alpha)A_i + C_i + M_{i+1} - M_i - \tau \tag{7.13}$$

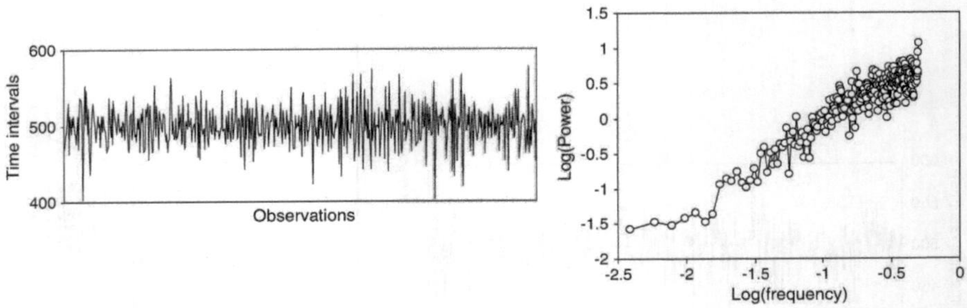

Fig. 7.8 *Right*: An example time series of inter-tap intervals collected in a synchronized finger tapping task. *Left*: Averaged log–log power spectrum. ($N = 13$; data from Torre and Delignières 2008)

In this model τ represents the fixed period of the metronome, and C_i and M_i are the components of the aforementioned Wing–Kristofferson model (internal timekeeper and motor delays, respectively). A_i represents asynchronies to the metronome and the model incorporates an autoregressive correction process with parameter $(1-\alpha)$. Time intervals are derived from a linear combination of successive asynchronies (Eq. 7.14).

$$I_i = A_i - A_{i-1} + \tau \tag{7.14}$$

In the initial formulation of the model, the timekeeper (C_i) was considered a white noise source (Vorberg and Wing 1996), but this hypothesis was unable to explain the pattern of correlations observed in interval series. Torre and Delignières (2008) showed that providing the timekeeper with *1/f* properties allowed satisfactorily reproducing the experimental results.

These two examples show that even in very simple laboratory tasks, LRC do not appear in isolation in experimental series and could sometimes be difficultly discernible. This is important because apparent alterations in long-range correlations could in fact be due to the superimposition of short-term correlations.

An interesting example has been described in the domain of walking. Hausdorff et al. (1996) showed that the series of stride intervals during walking contained LRC. However, they suggested that correlations tended to disappear when participants had to walk in synchrony with a metronome. The authors applied DFA of stride interval series and obtained α exponents close to one in self-paced walking, but close to 0.5 in metronomic walking. They concluded that “supraspinal influences could override the normally present long-range correlations” (p. 1456). In other words, the intention to walk in synchrony with the metronome was supposed to induce a kind of loss of complexity in the locomotor system, walking pace being externally regulated by the metronome. However, Delignières and Torre (2009), in a reassessment of the original data of Hausdorff et al., showed that the series of stride intervals, when participants had to walk in synchronization with a metronome, were not uncorrelated as postulated by the authors, but presented a rather complex pattern of correlation.

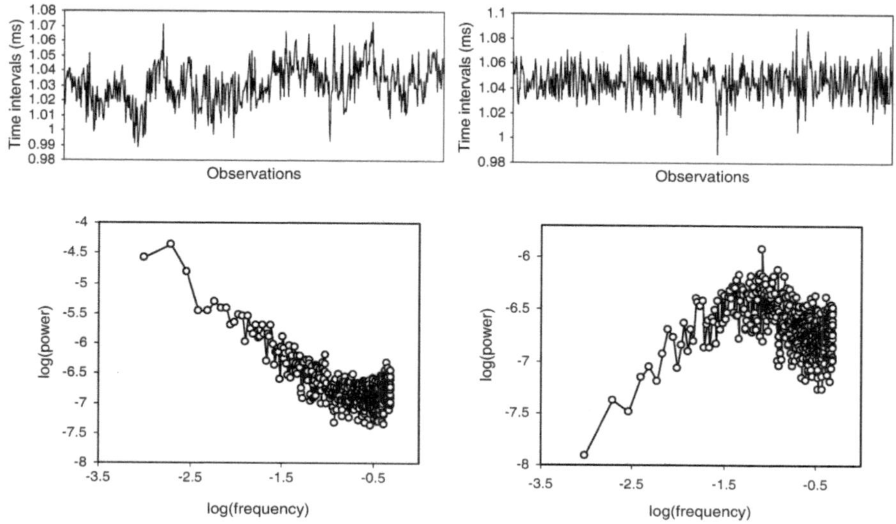

Fig. 7.9 Analysis of serial correlations in stride interval series during self-paced walking (*left column*) and during walking in synchrony with a metronome (*right column*). *Top row*: Example time series, *bottom row*: Averaged log–log power spectra. ($N = 10$, from Delignières and Torre 2009)

Especially the application of PSD revealed a log–log power spectrum with positive slope in low frequencies, and a slightly negative slope in the high frequency region (Fig. 7.9).

Delignières and Torre (2009) tried to account for this result with a model initially proposed by West and Scafetta (2003) and composed of a forced van der Pol oscillator:

$$\ddot{x} = \gamma\dot{x} - \lambda\dot{x}x^2 - \omega_i{}^2x + A\sin(\omega_0) + \sqrt{Q}\xi_t \quad (7.15)$$

This model is supposed to represent the dynamics of an oscillator, driven by a periodic forcing function. In this equation, the stiffness parameter (ω_i) is cycle-dependent and possesses fractal properties. The forcing term $\sin(\omega_0)$ represents the metronome with a fixed frequency ω_0, and A represents the coupling strength. The authors showed that setting A to 1 for self-paced walking and to 10 for synchronized walking allowed reproducing the patterns of correlations experimentally observed. These results show that during synchronization, LRC are still present in the system, but that the process of synchronization induces a short-term corrective process that could hide their presence.

Another example concerns finger tapping. Chen et al. (2001) showed that in synchronized finger tapping, the series of asynchronies to the metronome presented LRC. They also showed that the strength of LRC tended to increase when taps had to be performed in syncopation (i.e., in between metronome signals). The application of PSD revealed a steeper negative slope for syncopation than for synchronization (Fig. 7.10). The authors attributed the strengthening of LRC to the increased task

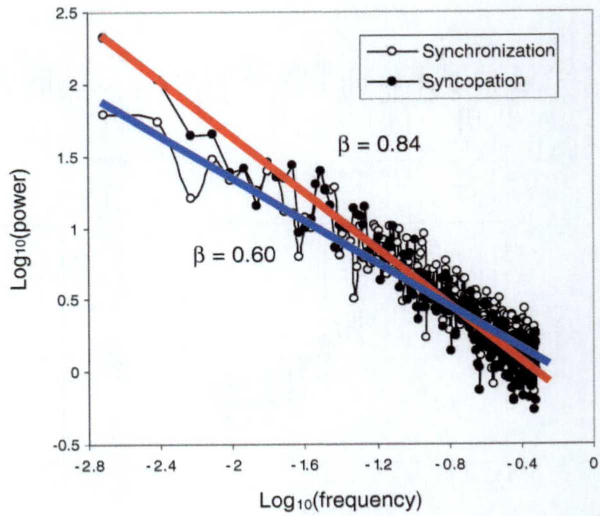

Fig. 7.10 Averaged log–log power spectra for series of asynchronies obtained during tapping in synchronization and in syncopation with a metronome. ($N = 11$; from Delignières et al. 2009)

difficulty of tapping between the beats. In terms of system complexity, task difficulty could be supposed to solicit more complex coordination between system's components, yielding a more robust *1/f* scaling.

However, the relationship between task difficulty and LRC is not so univocal: Sometimes an increased difficulty induces a strengthening of LRC, but in other experiments the opposite result can be obtained (Kello et al. 2007). Delignières et al. (2009) tried to provide an alternative explanation for the increase of LRC in syncopation tapping. They proposed to account for asynchronies during tapping in synchronization by the previously presented model (Eq. 7.13), derived from an original model by Vorberg and Wing (1996) and enriched by providing the timekeeper with *1/f* properties.

The authors proposed that in syncopation conditions, another process should be added in the model for the necessary estimation of half-period of the metronome, and they suggested that this process could be considered correlated over time (for further details see Delignières et al. 2009). Simulation results showed that the combination of this short-range correlated process with the previous model induced an increase of correlations in the resulting series, similar to the one empirically observed.

In sum, these experiments suggest that short-range processes can either decrease or conversely increase correlations in the signal. Fractal analysis methods are often unable to discern and identify these short-range processes, and can often yield erroneous interpretations. The application of different methods—in the time domain and in the frequency domain—and additional studies based on modeling and simulation are generally necessary for identifying the multiple sources of serial correlation that combine in experimental series.

False Detection of Long-range Correlations

An additional problem is that short-range correlation processes can sometimes mimic long-range correlations. In order to illustrate the possible misinterpretations that could arise with classical methods such as DFA or PSD, we simulated three time series, possessing distinct correlation properties. The first series was generated by a one-order autoregressive model:

$$y_i = \varphi y_{i-1} + \varepsilon_i \tag{7.16}$$

In this equation, φ is the autoregressive parameter and was set to 0.85. ε_i is a white noise process with zero mean and unit variance. The second series was generated by an integrated one-order moving average model:

$$y_i = y_{i-1} - \theta\varepsilon_{i-1} + \varepsilon_i \tag{7.17}$$

In this equation, θ is the first-order moving average parameter and was set to 0.8. ε_i is a white noise process with zero mean and unit variance. Finally, we used the Harte-Davies algorithm for simulating a series of fractional Gaussian noise, with $H = 0.9$ (Davies and Harte 1987). The three series are presented in Fig. 7.11. By construction, the two first series present only short-term correlations, while the third one presents LRC.

The corresponding DFA diffusion plots are presented in the *middle row* of Fig. 7.11. As can be seen in all cases, the diffusion plot presents a linear slope close to 0.9. Obviously the best linear fit is observed for the fGn series, which contains genuine long-range correlations. However the diffusion plots obtained for auto-regressive and moving average series mimic the typical shape expected from long-range correlated series, and could lead to erroneous interpretations.

Spectral analysis yields less ambiguous results (see Fig. 7.11, *bottom panels*): for the auto-regressive series, the log–log spectrum presents a typical flattening in low frequencies, revealing the absence of correlation beyond a given time lag. The log–log spectrum of the integrated moving average series presents a flattening in high frequencies due to the strong contribution of white noise terms in the generation of the process. In contrast, the spectrum of the fGn series exhibits a perfect linear slope over the whole range of frequencies. However, each spectrum presents a negative linear slope in a broad range of frequencies, and could induce an erroneous detection of LRC. Especially in the case of auto-regressive series, the flattening is often restricted to few points in the low frequency region, and the spectrum almost perfectly mimics a $1/f$ shape (Wagenmakers et al. 2004).

Evidencing the Presence of Long-range Correlation

The main problem with graphical methods such as PSD or DFA is that these methods cannot provide a statistical evidence for the effective presence of LRC in a series. Sometimes surrogate tests comparing the result obtained with an experimental series

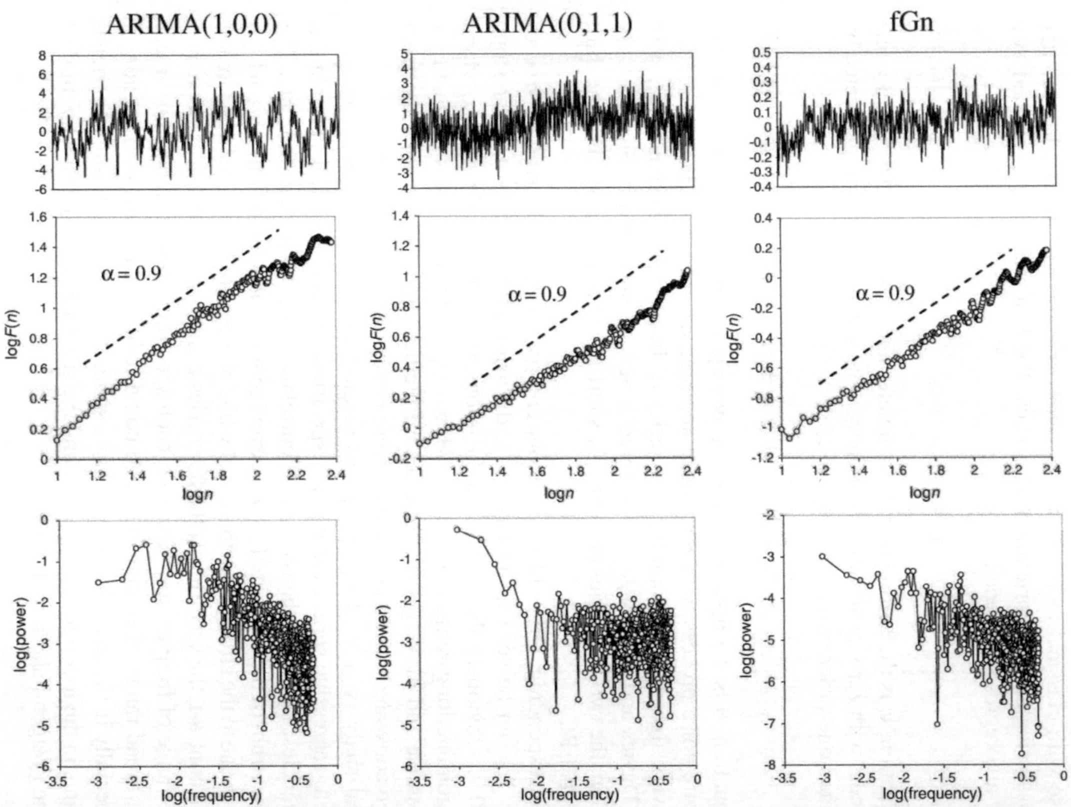

Fig. 7.11 Example series simulated with a one-order autoregressive model ($y_i = 0.85y_{i-1} + \varepsilon_i$, *left column*), a one-order moving average model ($y_i = y_{i-1} - 0.8\varepsilon_{i-1} + \varepsilon_i$, *central column*), and the Davies–Harte algorithm (fGn with $H = 0.9$, *right column*). The corresponding DFA diffusion plots are represented in the *median row*, and the log–log power spectra in the *bottom row*

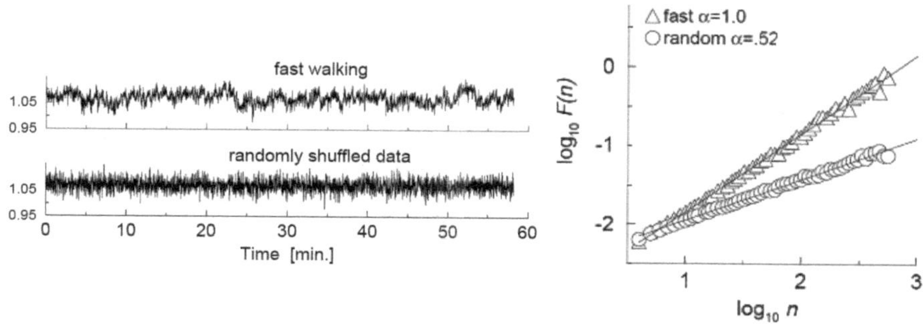

Fig. 7.12 Surrogate tests. *Left*: an example walking stride interval series *(top)* and a randomly shuffled version *(bottom)*. *Right*: DFA diffusion plots of the original and the shuffled series. The original series exhibits a $1/f$–like behavior, while the shuffled series yields a slope close to 0.5. (Hausdorff et al. 1996)

with those obtained from a set of shuffled versions of the original series are performed (Fig. 7.12). In such cases, the null hypothesis that is tested is the absence of correlation in the series. The test can eventually show that the series contains correlations, but cannot give any information about the (short- or long-range) nature of these correlations.

However, as stated by Slifkin and Newell (1998), the presence of correlations in empirical time series is more the rule than the exception. As such, the proper null hypothesis for testing the presence of LRC is not the absence of serial correlation but the short-range nature of correlations in the series.

Some specific methods have been developed for testing for the effective presence of LRC. In particular, Wagenmakers et al. (2005) and Torre et al. (2007b) have proposed a method based on ARMA and ARFIMA modeling.

ARMA models have been introduced by Box and Jenkins (1976). An ARMA model is the combination of autoregressive and moving average processes, and is noted (p, q), in which p is the order of the auto-regressive process, and q the order of the moving average process. A more complex class of models (ARIMA) is characterized by the inclusion of an integration process allowing to represent trends in the series. An ARIMA model is noted (p, d, q), d representing the order of the integration process. In the original formulation, p, d, and q are integers. Granger and Joyeux (1980) proposed to allow the integrating parameter d to take on fractional values. The resulting ARFIMA models (for *auto-regressive-fractionally integrated-moving average* models) present long-range correlation properties.

The method proposed by Wagenmakers et al. (2005) consists of fitting 18 models to the studied series. Nine of these models are ARMA (p, q) models, p and q varying systematically from 0 to 2. These ARMA models do not contain any long-range serial correlations. The other nine models are the corresponding ARFIMA (p, d, q) models, differing from the previous ARMA models by the inclusion of the fractional integration parameter d representing persistent serial correlations. One supposes that if the analyzed series contains LRC, the ARFIMA models should present a better fit

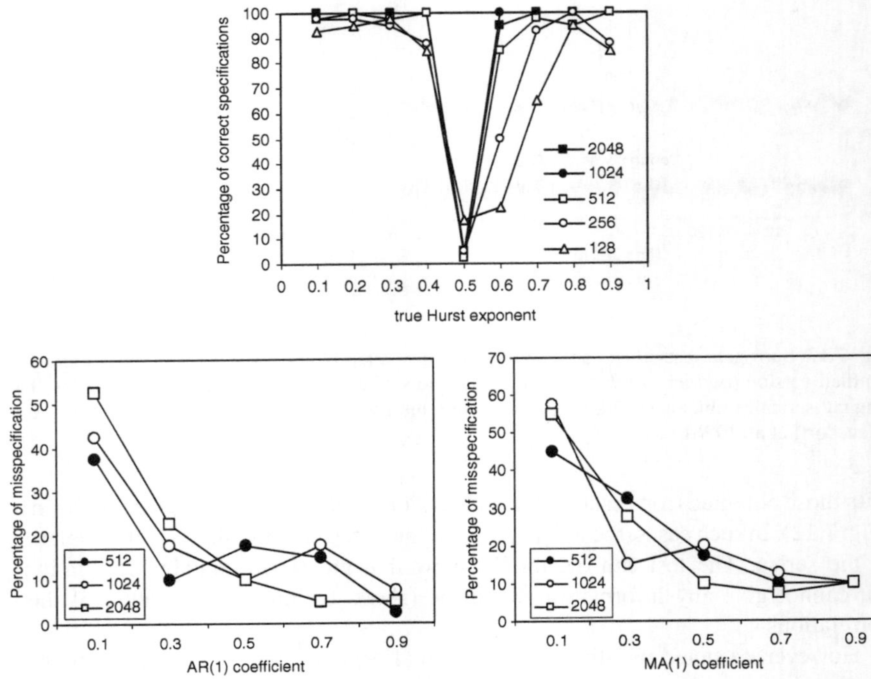

Fig. 7.13 ARMA/ARFIMA modeling. *Top panel:* percentage of correct specifications of exact fGn series (from $H = 0.1$ to $H = 0.9$), for different series lengths (from $N = 128$ to $N = 2048$). The percentage of correct specifications exceeds 90 %, except for very short, persistent series ($H > 0.5$, $N = 128$ and $N = 256$). *Bottom panels:* percentage of misspecifications for auto-regressive series (*left*) and moving average series (*right*), according to series length (from $N = 512$ to $N = 2048$) and to the value of the auto-regressive (AR(1)) and the moving average (MA(1)) parameters. The percentage of misspecifications remains moderate, except for very low values of AR(1) and MA(1). (From Torre et al. 2007b)

than their ARMA counterparts. The method selects the best model on the basis of a goodness-of-fit criterion (Bayes Information Criterion) based on a trade-off between accuracy and parsimony.

Torre et al. (2007b) proposed two complementary statistical criteria for ascertaining the presence of LRC: (1) The percentage of series better fitted by an ARFIMA rather than by an ARMA model, and (2) the mean sum of weights captured by ARFIMA models (the weight of a model is derived from the value of the goodness-of-fit criterion and represents the probability that this model is the best among the 18 candidate models for a given series). The authors proposed to accept the LRC hypothesis if at least 90 % of the series are best fitted by an ARFIMA model, and if the mean sum of ARFIMA weights exceeds 0.90 (note that the total sum of weights for the 18 models is 1). The authors showed that this method allowed to correctly detect LRC in simulated fGn series (see Fig. 7.13, *top panel*). They also showed that the percentage of false detections, in pure autoregressive and moving average series, remained moderate, except for very low values of p and q (see Fig. 7.13, *bottom panels*).

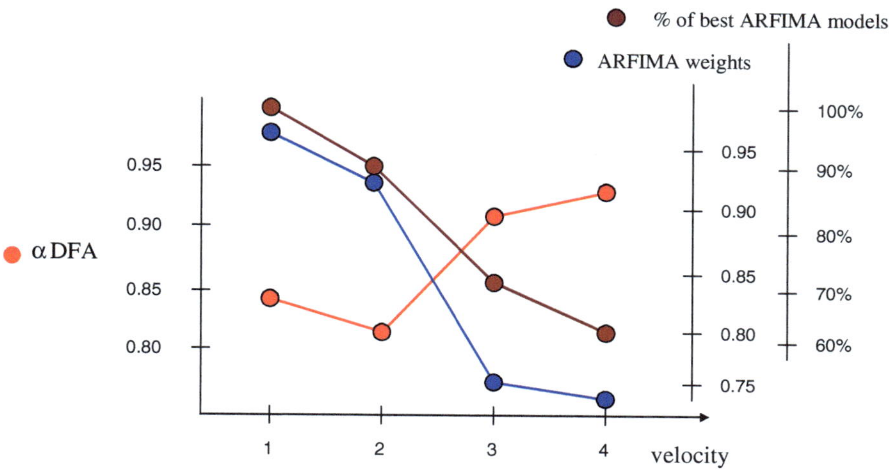

Fig. 7.14 Analysis of stride interval series during walking on a treadmill. Effect of walking velocity on the scaling exponent (αDFA), the sum of ARFIMA weights, and the percentage of series best fitted by ARFIMA models

Thornton and Gilden (2005) proposed an alternative method, a spectral likelihood classifier that used the shape of the power spectrum to decide among short- and long-range descriptions of data. Farrell et al. (2006) showed that the two methods gave roughly equivalent results.

As an example, we would like to present the results of an experiment during which participants had to walk on a treadmill at four different velocities (Marmelat and Delignières 2011). These four velocities were individually determined on the basis of a preliminary estimation of preferred velocity (the most comfortable and efficient velocity) and critical velocity (the velocity that induced a spontaneous transition towards running). At 80 % of preferred velocity v_1 was fixed, v_2 corresponded to preferred velocity, v_3 was fixed at equal distance between preferred and critical velocities, and v_4 was the critical velocity. We collected 512 stride intervals in each condition, and applied DFA and ARMA/ARFIMA modeling on the series.

Detrended Fluctuation Analysis revealed an effect of walking velocity, with a gradual increase of serial correlations with increasing velocity (Fig. 7.14), that could lead to a first conclusion: LRC tend to increase at high velocities. This first conclusion is consistent with those proposed in some previous studies (Hausdorff et al. 1996; Jordan et al. 2006). However, the application of ARMA/ARFIMA modeling showed that the global likelihood of ARFIMA models exceeded the threshold of 90 % and that the number of series best fitted by ARFIMA models reached 90 % only at slow and preferred speeds (Fig. 7.14). This yields a completely different conclusion: LRC tend to disappear at high velocities. This example clearly shows that the strength of serial correlation is not predictive of the genuine presence of LRC. Sometimes systems can present moderate levels of *effective* LRC, whereas in others cases, series can present high correlation levels but are not long-range correlated. Measuring strong correlations in a series does not guarantee that these correlations are long-range in nature.

Conclusion

The concept of LRC is currently particularly appealing for researchers. This approach allows to reveal new properties of the systems under study and to question traditional theories (Torre and Wagenmakers 2009). Especially the monofractal framework, considering $1/f$ fluctuations as an optimal compromise between order and disorder, suggests the adoption of new points of view concerning various key topics, including health (Goldberger et al. 2002a), adaptability (Harbourne and Stergiou 2009), and learning (Wijnants et al. 2009). Learning, for example, is generally conceived as the progressive installation of order in the system: with practice, trajectories become smoother, inter-trial variability decreases, etc. Wijnants et al. (2009) analyzed the effects of practice on LRC in the series of movement times in a cyclical aiming task. Their results showed that LRC tended to progressively increase and reach levels close to $1/f$ noise after a sufficient amount of practice. This suggests that learning could be essentially conceived as an optimization of interaction dynamics within the system, and explains why LRC are generally evidenced in series produced in usually practiced and overlearned tasks.

Long-range correlation analyses could appear easy to apply and their results easy to interpret in terms of deviation from the optimal $1/f$ noise. The examples developed in the present paper suggest, however, that the results provided by these analyses should be interpreted with caution. Especially $1/f$ fluctuations appear rarely in isolation in experimental series and could be contaminated by various short-range correlated processes. The resultant correlation pattern is often complex and can lead to erroneous conclusions.

References

Bassingthwaighte JB, Liebovitch LS, West BJ (1994) Fractal physiology. Oxford University Press, New York

Beltz BB, Kello CT (2006) On the intrinsic fluctuations of human behavior. In: Vanchevsky M (ed) Focus on Cognitive psychology research. Nova Science, Hauppauge, pp 25–41

Box GEP, Jenkins GM (1976) Time series analysis: forecasting and control. Holden–Day, Oakland

Caccia DC, Percival D, Cannon MJ, Raymond G, Bassingthwaigthe JB (1997) Analyzing exact fractal time series: evaluating dispersional analysis and rescaled range methods. Physica A 246:609–632

Cannon MJ, Percival DB, Caccia DC, Raymond GM, Bassingthwaighte JB (1997) Evaluating scaled windowed variance methods for estimating the Hurst coefficient of time series. Physica A 241:606–626

Chen Y, Ding M, Kelso JAS (2001) Origins of timing errors in human sensorimotor coordination. J Mot Behav 33:3–8

Davies RB, Harte DS (1987) Tests for Hurst effect. Biometrika 74:95–101

Delignières D, Fortes M, Ninot G (2004a) The fractal dynamics of self-esteem and physical self. Nonlinear Dynamics Psychol Life Sci 8:479–510

Delignières D, Lemoine L, Torre K (2004b) Time intervals production in tapping and oscillatory motion. Hum Mov Sci 23:87–103

Delignières D, Ramdani S, Lemoine L, Torre K, Fortes M, Ninot G (2006) Fractal analysis for short time series: a reassessement of classical methods. J Math Psychol 50:525–544
Delignières D, Torre K (2009) Fractal dynamics of human gait: a reassessment of the 1996 data of Hausdorff et al. J Appl Physiol 106:1272–1279
Delignières D, Torre K, Lemoine L (2009) Long-range correlation in synchronization and syncopation tapping: a linear phase correction model. PLoS ONE 4(11):e7822
Delignières D, Marmelat V, Torre K (2011) Paper presented to the skills conference 2011, Montpellier, December 15–16, 2011
Diniz A, Wijnants ML, Torre K, Barreiros J, Crato N, Bosman AMT, Hasselman F, Cox RFA, Van Orden GC, Delignières D (2011) Contemporary theories of 1/f noise in motor control. Hum Mov Sci 30:889–905
Einstein A (1905) Über die von der molekularkinetischen Theorie der Wärme geforderte Bewegung von in ruhenden Flüssigkeiten suspendieren Teilchen. Annalen Physik 322:549–560
Eke A, Herman P, Bassingthwaighte JB, Raymond GM, Percival DB, Cannon M, Balla I, Ikrényi C (2000) Physiological time series: distinguishing fractal noises from motions. Pflügers Arch 439:403–415
Eke A, Hermann P, Kocsis L, Kozak LR (2002) Fractal characterization of complexity in temporal physiological signals. Physiol Meas 23:R1-R38
Farrell S, Wagenmakers E-J, Ratcliff R (2006) Methods for detecting $1/f$ noise. University of Bristol, Bristol. http://seis.bris.ac.uk/~pssaf/tgreplysims.pdf. Accessed 3 Sept 2011
Gilden DL (1997) Fluctuations in the time required for elementary decisions. Psychol Sci 8:296–301
Gilden DL, Thornton T, Mallon MW (1995) $1/f$ noise in human cognition. Science 267:1837–1839
Goldberger AL, Amaral LAN, Hausdorff JM, Ivanov PCh, Peng C-K, Stanley HE (2002a) Fractal dynamics in physiology: alterations with disease and aging. PNAS 99:2466–2472
Goldberger AL, Peng C-K, Lipsitz LA (2002b) What is physiologic complexity and how does it change with aging and disease? Neurobiol Aging 23:23–26
Granger CWJ, Joyeux R (1980) An introduction to long-memory models and fractional differencing. J Time Ser Anal 1:15–29
Harbourne RT, Stergiou N (2009) Movement variability and the use of nonlinear tools: principles to guide physical therapist practice. Phys Ther 89:267–282
Hausdorff JM (2005) Gait variability: methods, modeling and meaning. J NeuroEng Rehabil 2:19. doi:10.1186/1743-0003-2-19
Hausdorff JM, Peng CK, Ladin Z, Wei JY, Goldberger AR (1995) Is walking a random walk? Evidence for long-range correlations in stride interval of human gait. J Appl Physiol 78:349–358
Hausdorff JM, Purdon PL, Peng CK, Ladin Z, Wei JY, Goldberger AR (1996) Fractal dynamics of human gait: stability of long-range correlations in stride interval fluctuation. J Appl Physiol 80:1448–1457
Hausdorff JM, Mitchell SL, Firtion R, Peng CK, Cudkowicz ME, Wei JY, Goldberger AL (1997) Altered fractal dynamics of gait: reduced stride-interval correlations with aging and Huntington's disease. J Appl Physiol 82:262–269
Hurst HE (1951) Long-term storage capacity of reservoirs. T Am Soc Civ Eng 116:770–799
Hurst HE (1965) Long-term storage: an experimental study. Constable, London
Ihlen EAF, Vereijken B (2010) Beyond $1/f^{\alpha}$ fluctuation in cognitive performance. J Exp Psychol Gen 139:436–463
Jordan K, Challis JH, Newell KM (2006) Long range correlations in the stride interval of running. Gait Posture 24:120–125
Kello CT, Beltz BC, Holden JG, Van Orden GC (2007) The emergent coordination of cognitive function. J Exp Psychol Gen 136:551–568
Kello CT, Brown GDA, Ferrer-i-Cancho R, Holden JG, Linkenkaer-Hansen K, Rhodes T, Van Orden G (2010) Scaling laws in cognitive sciences. Trends Cogn Sci 14:223–232
Leland WE, Taqqu MS, Willinger W, Wilson DV (1994) On the self-similar nature of Ethernet traffic (extended version). IEEE/ACM Trans Netw 2:1–15

Lemoine L, Torre K, Delignières D (2006) Testing for the presence of 1/*f* noise in continuation tapping data. Can J Exp Psychol 60:247–257

Mandelbrot BB, Van Ness JW (1968) Fractional Brownian motions, fractional noises and applications. SIAM Rev 10:422–437

Marmelat V, Delignières D (2011) Complexity, coordination and health: avoiding pitfalls and erroneous interpretations in fractal analyses. Med Lith 47:393–398

Matsuzaki M (1994) Fractals in earthquakes. Philos Trans: Phys Sci Eng 348:449–457

Peng C-K, Mietus J, Hausdorff JM, Havlin S, Stanley HE, Goldberger AL (1993) Long-range anti-correlations and non-Gaussian behavior of the heartbeat. Phys Rev Lett 70:1343–1346

Pressing J, Jolley-Rogers G (1997) Spectral properties of human cognition and skill. Biol Cybern 76:339–347

Slifkin AB, Newell KM (1998) Is variability in human performance a reflection of system noise? Curr Dir Psychol Sci 7:170–177

Thornton TL, Gilden DL (2005) Provenance of correlations in psychophysical data. Psychon Bull Rev 12:409–441

Torre K, Delignières D (2008) Unraveling the finding of 1/*f* β noise in self-paced and synchronized tapping: a unifying mechanistic model. Biol Cybern 99:159–170

Torre K, Wagenmakers EJ (2009) Theories and models for 1/f noise in human movement science. Hum Mov Sci 28:297–318

Torre K, Delignières D, Lemoine L (2007a) $1/f^{\beta}$ fluctuations in bimanual coordination: an additional challenge for modeling. Exp Brain Res 183:225–234

Torre K, Delignières D, Lemoine L (2007b) Detection of long-range dependence and estimation of fractal exponents through ARFIMA modeling. Br J Math Stat Psychol 60:85–106

Van Orden GC (2007) The fractal picture of health and wellbeing. Psychol Sci Agenda 21(2). http://www.apa.org/science/psa/vanorden.html. Accessed 1 July 2009

Van Orden GC, Holden JC, Turvey MT (2003) Self-organization of cognitive performance. J Exp Psychol Gen 132:331–350

Vorberg D, Wing A (1996) Modeling variability and dependence in timing. In: Heuer H, Keele SW (eds) Handbook of perception and action, vol 2. Academic Press, London, pp 181–262

Wagenmakers E-J, Farrell S, Ratcliff R (2004) Estimation and interpretation of $1/f^{\alpha}$ noise in human cognition. Psychon Bull Rev 11:579–615

Wagenmakers E-J, Farrell S, Ratcliff R (2005) Human cognition and a pile of sand: a discussion on serial correlations and self-organized criticality. J Exp Psychol Gen 134:108–116

West BJ (1990) Fractal physiology and chaos in medicine. World Scientific, Singapore

West BJ, Scafetta N (2003) Nonlinear dynamical model of human gait. Physical Review E 67:051917

Wijnants ML, Bosman AMT, Hasselman F, Cox RFA, Van Orden GC (2009) 1/*f* scaling in movement time changes with practice in precision aiming. Nonlinear Dyn Psychol Life Sci 13:79–98

Wing AM, Kristofferson AB (1973) The timing of interresponse intervals. Percept Psychophys 13:455–460

Chapter 8
On the Control of Unstable Objects: The Dynamics of Human Stick Balancing

Ramesh Balasubramaniam

Introduction

Objects that we control and interact with are often unstable. Riding a bicycle, balancing a tray of food, maintaining the oscillations of a hula-hoop, and even standing upright are exemplary tasks that require the control of an unstable object. Although it is difficult to characterize the physics of complex object interactions, we are adept at learning and performing these tasks in everyday life. Unstable objects require carefully assembled control mechanisms because, by definition, the object must be stabilized through the interaction between the human control and the intrinsic object dynamics. Additionally, such tasks demand extremely precise control because error can elicit abrupt and irrevocable changes in the performance (Balasubramaniam and Turvey 2004; Cluff et al. 2008).

Although much is known about human motor control and object manipulation in predictable systems (when the mapping between actions and their consequences is straightforward), much less is understood about unstable object control. This is largely because the dominant research focus has been on characterizing the task and context-dependent attributes of firmly grasped, rigid object control (Imamizu et al. 2003; Milner et al. 2006). Although an extensive literature has focused on adaptation to novel mechanical loads, few studies have considered how we learn to control unstable objects. As a result, there are a number of important questions that remain in motor control research: What strategies are used to control unstable objects and how are these control mechanisms learned? Are common control processes shared between interacting motor systems? Do these control mechanisms involve predictive internal forward models?

In this review article, I use a stick-balancing task as a paradigmatic example to investigate control mechanisms and skill acquisition in relation to unstable object control. The stick-balancing task imposes a complex control problem that involves

R. Balasubramaniam (✉)
Sensorimotor Neuroscience Laboratory, McMaster University,
Hamilton, ON L8S 4K1, Canada
e-mail: ramesh@mcmaster.ca

M. J. Richardson et al. (eds.), *Progress in Motor Control,*
Advances in Experimental Medicine and Biology 782,
DOI 10.1007/978-1-4614-5465-6_8,

maintaining an inverted pendulum in dynamic equilibrium at the finger tip. First, the number of degrees of freedom that needs to be controlled far exceeds the dimensions of the task. Although the stick moves freely in three spatial dimensions, a large number of body segments have been coordinated in order to keep the stick upright and stable. Second, and more important, as the stick is allowed to pivot freely, the effect of forces applied at the fingertip depends on the angular state of the stick (i.e., position and velocity). Small errors in the estimation and detection of state-specific information could translate to a serious loss of performance stability.

The inverted pendulum control problem has been the object of study in control systems engineering and human motor control (Narendra and Annaswamy 1986). Previous studies have generated a number of important insights on the dynamical and neural control processes involved (Treffner and Kelso 1999; Foo et al. 2000; Mah and Mussa-Ivaldi 2003a, b; Cabrera et al. 2006), which fall under the general classification of two theories: internal model and intermittent feedback control. In the following section, I review the two theoretical frameworks, as they pertain to stick balancing in detail, before describing recent work from our laboratory on this topic.

Issues in Studying Unstable Object Dynamics

Internal Forward Model-Based Control

A large body of evidence in human reaching tasks suggests that humans develop internal models to produce the systematic forces required to deal with force-field perturbations (Lackner and DiZio 1994; Shadmehr and Mussa-Ivaldi 1994. Following the seminal work of Wolpert et al. (1995), an important theoretical development is that the brain acquires and uses an internal model that encodes the physical properties of our limbs (Singh and Scott 2003; Kurtzer et al. 2008), environment, and manipulated objects. In particular, object manipulation tasks (Ahmed et al. 2008) have suggested the existence of an internal model that captures the relationship between forces applied to an object and its ensuing movement. Further, Mussa-Ivaldi and colleagues have shown that the relationship between applied force and motion can be learned in the absence of upper limb movement (Mah and Mussa-Ivaldi 2003a). Once such a model is acquired, it can be generalized to novel limb configurations, but such a model does not transfer to objects with different dynamics (Mah and Mussa-Ivaldi 2003b). The general conclusion from these studies is that the control of objects requires knowledge of the physical properties of the object. In the context of stick balancing, these results imply that the successful control of an unstable stick requires object-specific knowledge which is acquired and maintained by the upper limb movements, the sensory consequences of these movements, and the resulting stick motion.

A question of interest in this review is: are internal models required to balance an inverted pendulum? The internal model approach would predict that the stick-balancing task could be performed by predicting sensory information about the inverted pendulum's angular position and velocity. In much of the recent work on internal models in motor control, a major theoretical point has been that sensory signals that provide state information about the stick (position and velocity) are influenced by noise (for review, see Faisal et al. 2008) and time delay. In the wake of sensory uncertainty and time delay, it has been argued that internal models can be used to estimate the state of the body, environment, and manipulated objects. Such models are acquired and maintained by combining efference copies of motor commands and sensory feedback of the movements to predict the sensory consequences of movements (Wolpert et al. 1998). State estimates are generated using principles to minimize uncertainty via a Kalman filter (Kalman 1960) that uses internal feedback based on the motor command and sensory feedback in conjunction with a model of the motor system.

In a recent study, Mehta and Schaal (2002) examined internal models in the visuomotor control of stick balancing. They found that subjects could successfully balance a stick even in the absence of visual information (in blank-out trials lasting up to 600 ms) and in the absence of force feedback about the stick's state. They concluded that the central nervous system (CNS) uses a forward model to control the stick, but were unable to show the form of model-based control used in stick balancing (Mehta and Schaal 2002). It is possible that this was due to the limitations of a Kalman filter based state estimation mechanism for handling the type of statistical distributions seen in stick balancing (Cluff and Balasubramaniam 2009). Although I do not take up the issue of state estimation directly in this review, I highlight a few caveats about using internal models on the basis of Kalman filter based observer models.

Intermittency and Dynamical Systems Accounts of Stick Balancing

An alternative approach to the internal model account has been developed by Milton et al. (2009). According to these authors, continuous balance control does not adequately describe the behavioral strategies used to control unstable objects. They argue that the difficulty of controlling an inverted pendulum arises due to limitations in simultaneously processing noisy time-delayed feedback while specifying controlled motor responses (Milton et al. 2009). They have contested the viability of continuous model-based control following experimental evidence showing that intermittent rather than continuous control strategies are used in stick-balancing tasks in the context of feedback uncertainty and delay (Loram et al. 2006; Milton et al. 2009; Gawthrop et al. 2011).

In recent years, Cabrera and Milton (2002, 2004a) have shown that stick balancing shows characteristics of intermittent control. They observed that stick displacements exhibit alternating small and large amplitude excursions with frequency. Two important power-law relationships can be seen here. First, the power spectrum of stick fluctuations follows a $-1/2$ power law. When the laminar phases were analyzed, the distribution revealed a $-3/2$ power law. Cabrera and Milton (2002) observed that corrective stick movements were performed at all time scales; the modal ones occurring at time intervals that are shorter than sensorimotor delays in human voluntary movement ($\sim$100 ms). Evidence for intermittent control mechanisms have since been reported for the manual control of unstable virtual load (Loram et al. 2009). Further, Gawthrop et al. 2011, have successfully modeled the intermittent control strategy employing ballistic control forces that operate when the angular deviations of a stick exceed specific thresholds. Such a discontinuous control mechanism reflects the usage of short time scale, stochastic forcing of objects when they cross set-point stability boundaries (Cabrera and Milton 2002, 2004a).

It is of interest to note that stick-balancing time is inversely related to the weight and height of the stick. Lighter and shorter sticks are more difficult to control. Periodic vibrations, even shaking an object with the other hand while balancing a stick, help to stabilize performance. These observations suggest that intermittent control could be related to feedback uncertainty, time delay, and interactions between the two. However, the adaptive nature of intermittent control has yet not been explored. It is likely that a combination of the stochastic processes underlying basic hand position and feedback control processes generate intermittency in stick-balancing control (Wolpert et al. 1992; Treffner and Kelso 1999). In this review, I focus on quantifying such intermittent dynamics in stick balancing and accompanying posture control mechanisms.

Task-Specific Control of Upright Posture

It is important to consider the context and task environment in stick balancing. The task of balancing the stick takes place against an almost constant backdrop of the control of upright stance. In the following, I describe recent developments in standing balance control that have contributed to our understanding of stick balancing and the control of unstable objects in general.

Upright posture is stabilized by activity in distributed muscle groups that is scaled to the magnitude and the direction of self-generated and environmental forces (Ting and Macpherson 2005). Despite the complexity of the neural mechanisms involved in postural control, the mechanical basis of standing balance is to maintain (the vertical projection of) the center of mass within a support surface. Posture control is typically studied using time-varying properties of the body's center of pressure (COP). To the extent this equilibrium requirement is satisfied, the postural system appears to be recruited to facilitate goal-directed behavior. An emerging argument, pioneered by Stoffregen and colleagues, is that the diversity of voluntary control is inseparable from the postural mechanisms that support behavior (Riccio and Stoffregen 1988; Marin et al. 1999; Stoffregen et al. 1999; Stoffregen et al. 2000). Recent studies have shown that postural sway helps to facilitate the performance of tasks that are superordinate to the task of maintaining upright balance. This has been demonstrated in the context of light touch (Riley et al. 1999) and precision aiming (Balasubramaniam et al. 2000).

The analysis of postural fluctuations has revealed the interplay between stochastic and closed-loop feedback correction processes, as seen in the multiscale analysis of stick balancing. Postural fluctuations have been modeled widely using a dual timescale model (Collins and DeLuca 1994, 1995; Zatsiorsky and Duarte 1999). However, standing balance is flexible enough to be entrained by an external stimulus (Marin et al. 1999) and robust in the context of visual feedback delays (Boulet et al. 2010). Other studies have revealed task- and context-specific control of posture (Jeka et al. 2000; Kiemel et al. 2002; Peterka and Loughlin 2004). Other findings, however, imply an intimate functional link between posture and upper limb control (Ahmed and Wolpert 2009) that extends beyond the instability of standing balance (Balasubramaniam and Wing 2002), suggesting that a common predictive mechanism might underlie the control of both systems (Flanagan and Wing 1997). In this review, I will look at the interaction between the postural system and stick-balancing dynamics. In particular, I will examine how task parameters influence the control system underlying the two.

Perspectives in Motor Learning

Stick balancing does not come easy. Learning to balance an unstable object requires mastery of the degrees of freedom of the body and an understanding of the physical dynamics of the object. In this section, I will review how scientists have approached the issue of skill acquisition in recent years. The motor learning literature may be divided into (at least) two distinct approaches: sensorimotor adaptation and coordination dynamics.

Sensorimotor adaptation paradigms have been used to study how motor commands are modified in the wake of changing environments (Lackner and DiZio 1994; Shadmehr and Mussa-Ivaldi 1994). In this approach, learning has been argued to reflect an optimal parameter estimation process that serves to reduce error. Anticipatory change in reaching kinematics has been documented extensively in force-field adaptation studies, leading to important discoveries about trial-to-trial learning, consolidation, and interference. Imaging studies have revealed that the cerebellum (Imamizu et al. 2003) and basal ganglia (Seidler et al. 2001) are strongly implicated in sensorimotor adaptation and the modification of motor commands in changing environments. Although the sensorimotor adaptation paradigms have contributed to our understanding of learning movement trajectories, there have been few studies that have investigated the role of the interaction between the multielement structure of the body (and its many degrees of freedom) and various subsystems as a function of learning.

The coordination dynamics perspective offers a powerful framework to investigate the organization, stability, and control of voluntary movement. The largest successes of this perspective have been in quantifying the acquisition of bimanual coordination patterns. In coordination dynamics, the focus of research has been on learning induced changes in the spatio-temporal properties of a system, characterized by an order parameter (Zanone and Kelso 1992). In this approach, changes in an abstract parameter are believed to reflect dynamical events unfolding at multiple time scales (such as relative phase between two limbs). Learning has been described by the evolution of the topological properties that characterize body segment relationships and systematic changes in the recruitment and patterning of multiple degrees of freedom (Vereijken et al. 1992).

Given the number of degrees of freedom that have to be coordinated in complex motor learning situations, interactions between motor subsystems have also been a key part in dynamical approaches to motor learning. Following the seminal work of Bernstein (1967), Newell et al. (2001) have suggested that motor learning is instantiated by the evolving coordination of interacting motor subsystems. Their model distinguishes between three hierarchical levels of the motor system ranging from (1) individual effectors that operate and evolve within effector systems (e.g., individual muscles, segments, or joints) to (2) coordinative relationships between motor subsystems (e.g., posture and upper limb movement patterns) that interact to produce and stabilize (3) outcome performance. Newell et al. (2001) have proposed that such specialized motor subsystems are organized hierarchically. Such subsystems are assembled into functional units that support outcome performance. In this review, I will employ the ideas that have evolved from the motor learning literature, specifically related to the spatio-temporal properties and control of degrees of freedom, to describe skill acquisition in stick balancing.

Issues at Hand for Stick Balancing

It is, thus, evident that task- and context-specific control mechanisms are common features of posture and upper limb control. I posit that object manipulation skills in the task of balancing an unstable object, like a stick at the fingertip, are established through these common learning and control mechanisms governing these subsystems. I ask the following four questions. (1) How do we learn to control unstable objects? (2) Are stick-balancing dynamics intermittent and if so how can we quantify this intermittency? (3) Are distinct motor systems, such as the control of individual joints, posture, and the upper limb, linked through common learning and control processes? (4) What is the role of higher cognitive and attentional processes in the acquisition and maintenance of the stick-balancing skill?

Spatio-temporal Dynamics of Stick Balancing

The objective of our first study (Cluff and Balasubramaniam 2009) was to determine if the power-law scaling in stick balancing (Cabrera and Milton 2004a, b), described in the section above, is dependent on motor learning.

In order to test this idea, participants balanced a wooden dowel with length 62 cm, diameter 0.635 cm, and mass 50 g in two experimental conditions: sitting and standing. Sitting trials were performed with subjects seated comfortably in a chair at the subjects' preferred seat height. The subjects were required to balance the pole with their back remaining in contact with the seat. In the standing condition, subjects performed pole balancing with their feet approximately shoulder-width apart, but were able to move their upper body while maintaining a stationary stance.

Participants learned to balance a small cylindrical stick on their fingertip over a 2-week period. Data collection was performed on the first day, followed by subsequent data collection every fourth day. The subjects performed 30 min of daily practice between data collection sessions (15 min per condition), which was distributed between conditions according to their preference. We avoided confounding learning effects by counterbalancing the order of conditions across subjects. Motion capture was performed with eight VICON MX-40+ infrared cameras sampled at 500 Hz (Denver, CO, USA). We tracked pole motion in three-dimensions using two markers affixed to the top and bottom on the pole.

We computed the Euclidean finger tip speed $s(t)$ and the detrended speed, Δs. For the probability of a given step size, we computed $P(\Delta s, \Delta t)$ by plotting histograms with bin size set to 1 mm/s. To determine whether the probability of a given step size was influenced by the time between observations, Δt, we decimated $\Delta s(t)$ on a logarithmic scale by factors 1–1,000. We plotted the probability of return (i.e., the probability of zero change in fingertip speed between observations), $P(0, \Delta t)$, as a function of Δt. The power-law exponent α was computed by regressing $P(0, \Delta t)$ onto Δt on a log-log scale. Figure 8.1a shows the distribution of $P(\Delta s, \Delta t)$ for the sitting and the standing conditions. The purposes of this study were two-fold: first, to determine whether the decay exponent for the probability of a given step size, α, changed with learning, and second, to determine whether α varied in a sitting versus standing condition.

As subjects learned the mean balancing time increased, unequivocally suggesting that they were getting better at the task. Figure 8.1b shows the change in the value of the power-law exponent across sessions of learning. A careful look at Fig. 8.1 reveals that the values of α range from 0 to 2 (i.e., $0 < \alpha < 2$), suggesting that stick-balancing dynamics are Lévy-distributed, a finding previously reported by Milton et al. (2004). The Lévy process can be characterized as an unbounded and unconstrained random walk. The unbounded, asymptotic character of the Lévy distribution results in an infinitely variant process, resulting in the absence of the first and second statistical moments. Figure 8.1 also reveals that the distribution broadens with learning, this corresponds to a smaller decay in the probability for large step sizes. Behaviorally, this is manifested as tolerance to stochastic processes: the participant becomes more tolerant to large changes in fingertip speed with increasing task proficiency.

Our results demonstrate that motor learning results in increased tolerance for large stick displacements. The decay exponent α was influenced by learning, becoming significantly smaller with experience and resulting in less severe decay in the probability for a given velocity step size, $P(\Delta s, \Delta t)$. Moreover, the decay exponent α for $P(\Delta s, \Delta t)$ was greater in a sitting versus standing condition. Our results show conspicuously that both decay exponents and truncation change with learning, resulting in an increased tolerance to large fingertip excursions in pole balancing.

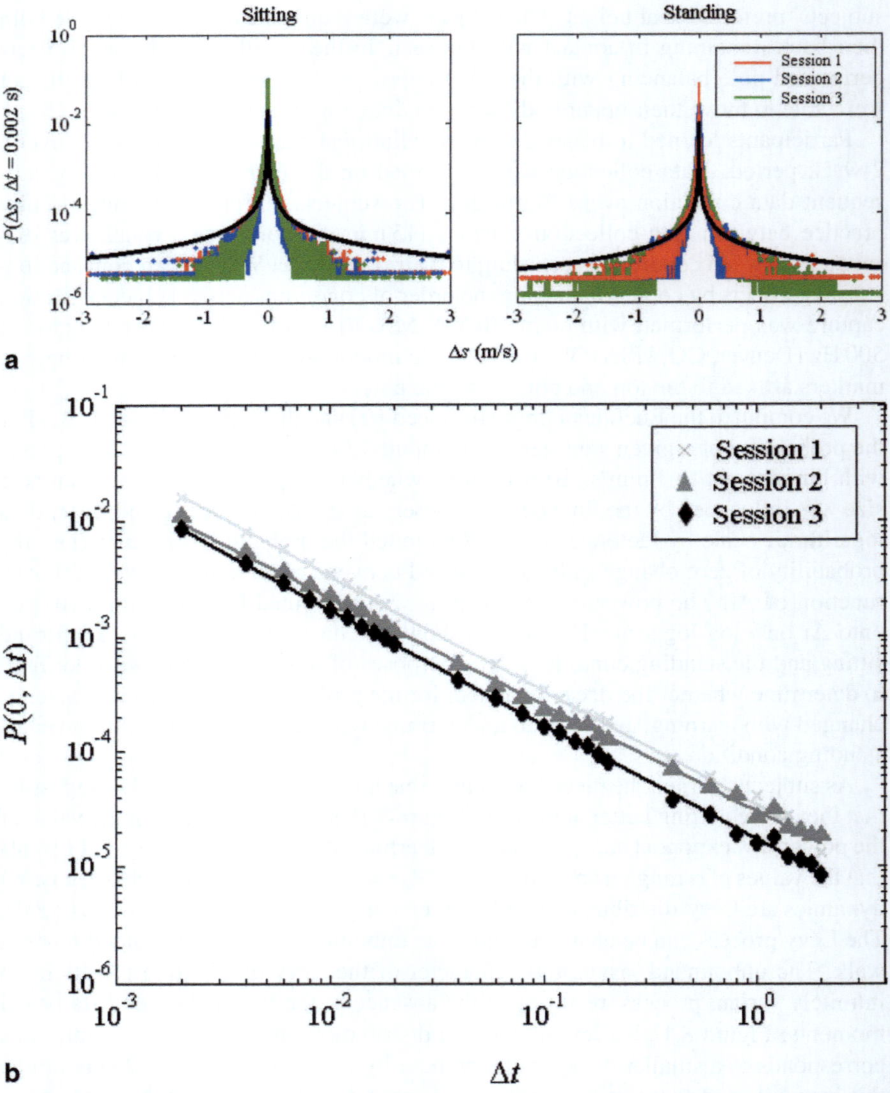

Fig. 8.1 *Top-panel:* Session 1; blue: session 2; green: session 3. Solid black line represents theoretical Lévy distribution with **a** $\alpha = 0.95$ and scale parameter $\gamma = 0.03$, **b** $\alpha = 0.98$ and scale parameter $\gamma = 0.025$. The overlaid theoretical Lévy distribution demonstrates both decay exponent α and truncation change with learning in the standing condition. *Bottom panel:* $P(0, \Delta t)$ follows a power-law distribution for $\Delta t = 0.002$ to 2 s, in the sitting condition. (Reprinted with permission from Cluff et al. 2009, Public Library of Science)

Our results that stick-balancing trajectories (probability of finger tip speed change over time interval) are Lévy-distributed, raises important concerns about hypothesized control mechanisms that are based on predictive internal models that employ

Kalman filters (Mehta and Schaal 2002). Lévy processes are indicative of nonpredictive search processes or foraging. Morover, a conventional Kalman filter assumes additive Gaussian processes and measurement noise. It is unlikely that a technique using conventional Kalman filters can be successfully used to model systems with multiplicative noise that yields power-law distributed variables.

In summary, this study has demonstrated that learning, which reflects changes between the dynamics of passive and predictive mechanisms, can be captured by changes in ensemble statistical distributions that capture the spatio-temporal properties in stick balancing.

Quantifying the Intermittency in Stick-Balancing Dynamics

In the previous section, I reported on the nature of Lévy distributions and power-law scaling seen in stick balancing. The question that remains is what kind of control mechanisms are implicated in seeing such a distribution. One interesting possibility that this observation raises is the presence of intermittent corrections at multiple time scales. Previous work by Milton et al. (2004) has shown that power-law scaling was also evident in the laminar phases (time intervals) for successive corrective movements, demonstrating that corrective movements were intermittent. In confirmation of intermittent control, behavioral data demonstrated that 98 % of corrective movements were shorter than our sensory processing delays ($\sim$100 ms). Numerical analyses have since demonstrated that balance is facilitated in time-delayed stochastic systems, provided the system is tuned near a stability boundary. In this case, control could result from stochastic processes that force the fingertip trajectory back and forth across stability boundaries. It is often argued that intermittent control might be favored to continuous estimation in stochastic, time-delayed systems as the computational burden incurred by the CNS is minimized (Milton et al. 2008).

The first goal of our next study (Cluff et al. 2009) was to perform a detailed investigation of the Lévy-distributed dynamics of stick-balancing fingertip trajectories and test for the presence of intermittent control mechanisms. To investigate this, we applied recurrence quantification analysis (RQA) to the fingertip displacement time series recorded during stick balancing (Webber and Zbilut 1992). An objective of this study was to quantify the intermittency seen in stick balancing and the changes accompanying learning. Intermittent systems are characterized by two distinct states, "off": a period over which dynamical variables are approximately constant and "on": where sudden, intermittent bursting of activity can be seen. In such systems, when the dynamical variable remains within a certain threshold bound it is quiescent. When the threshold bound is crossed, the system transitions from the "off" to "on" state, where a burst of activity might be seen.

The earlier section summarized the work that showed that individuals became tolerant of large amplitude fingertip displacements with pole balancing experience. This tolerance reflects an increased robustness to perturbations, a form of dynamical stability we sought to quantify. In addition, RQA provides a method for quantifying change in the degree of relative determinism versus stochasticity (*%DET*), robustness to perturbations (*L*max), nonstationarity (TREND) all embedded in pole-balancing fingertip dynamics—a characteristic of the dynamics that might be expected to change over the course of learning (cf. Riley et al. 1999; Balasubramaniam et al. 2000; Balasubramaniam and Turvey 2004). Finally, RQA provides a set of measures capable of indexing intermittency in the control enacted in pole balancing, including *%LAM*, *v*max, and TTIME.

Recall that in the earlier section, we reported significant differences in learning the stick-balancing task in the sitting and the standing conditions (Cluff and Balasubramaniam 2009). A key objective of this study was to assess the nature of the difference between the sitting and standing condition. We reasoned that the availability of greater number of biomechanical degrees of freedom in standing greatly contributed to the better acquisition of the stick-balancing skill in that condition. Earlier work has shown that learning can change the orderly recruitment of degrees of freedom into organized coordinative structures, in a manner that would facilitate performance (Bernstein 1967; Vereijken et al. 1992).

As with the previous study reported in the section, Spatio-temporal Dynamics of Stick Balancing, subjects learned to balance a stick in sitting and standing conditions. We analyzed the stick-balancing displacement time-series data using RQA. Results revealed a number of changes in the dynamics of fingertip displacements that occurred over the course of learning. RQA also revealed a number of effects related to the availability of biomechanical degrees of freedom for task performance. *%REC* is a measure of temporal correlation. It reflects the tendency for points that over time return to the same local neighborhood of the reconstructed phase space. *%REC* decreased progressively with learning, suggesting that temporal correlation in fingertip displacement series decreased with experience. Therefore, as participants became more experienced in balancing, the trajectories in the reconstructed phase space were less likely to repeat. Figure 8.2 summarizes the findings on recurrence rate as a function of learning for the sitting and the standing conditions.

Our results also demonstrated that Laminarity index (*%LAM*), and trapping time (TTIME,) which index intermittency in the dynamics, were all larger in the standing relative to sitting condition. Collectively, these results suggest that the underlying control was more intermittent for the standing condition. In other words, the system's propensity for intermittency was observed in relatively longer phases whereby the fingertip position was approximately constant. These results are consistent with a control mechanism that capitalizes on passive motor control dynamics and corrects for pole excursions only when these displacements threaten stability. Such a mechanism is often termed "drift and correct," following the work of Milton et al. (2004).

In summary, we have shown that learning resulted in greater stability of stick movement trajectories (resistance to perturbation); although they showed a greater tendency to return to the same areas of the reconstructed phase space. The availability of greater degrees of freedom in standing resulted in intermittent dynamics

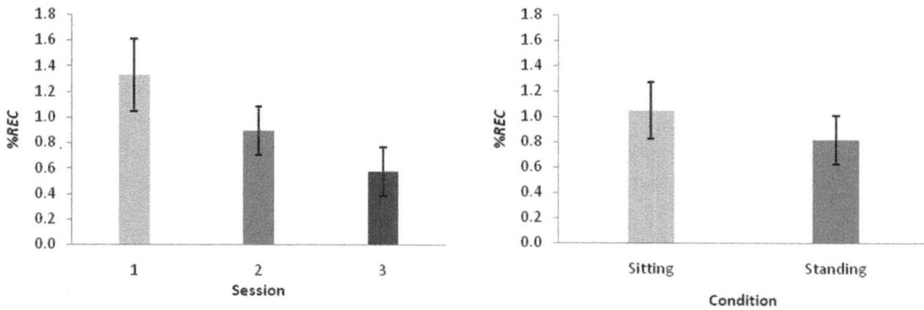

Fig. 8.2 *%REC* was moderated by learning and a condition effect. **a** %REC was dependent on a learning effect, decreasing progressively from the first through third experimental session. **b** %REC was dependent on a condition effect, with *%REC* greater in the sitting relative to standing condition, which reveals greater tendency for the dynamic to visit local neighborhoods in phase space in this condition. (Reprinted with permission from Cluff et al. 2009 @ Elsevier)

at the fingertip and suggested the role of the recruited "coordinative structures" in minimizing the computational burden on the CNS.

Task-Specific Coupling Between Posture and Hand

As mentioned in an earlier section, stick balancing takes place on top of the usual balancing and cognitive demands placed on the standing performer. Our results have also shown overwhelmingly that standing actively contributes to stick-balancing expertise and the intermittent control mechanisms that are characteristic of learned performance. Thus, a natural question to study would be: what are the interactions between the body's COP and stick trajectories when an actor learns the task of stick balancing. Cluff et al. (2011) examined this relationship and coupling between hand and postural displacements during stick balancing.

In this study, participants learned to balance a cylindrical wooden stick on their index finger while standing in an upright posture. Learning was quantified over four experimental sessions. Data collection sessions took place every fifth day and were about 90 min in duration. Subjects performed 30 min of daily practice between experimental sessions. RQA analyses were performed separately on hand and postural displacements. Cross-RQA (CRQA) was performed to study the co-time evolution and relationship between postural and hand displacements as subjects learned the task. We hypothesized, following the work of Newell et al. (2001) that we reviewed earlier, that learning would involve a reorganization of postural control to support stick performance.

In line with our previous studies, finger trajectories became more discontinuous with learning. But we also noted similar changes in the COP time series, suggesting that postural sway was indeed facilitating the performance of the "suprapostural" task of stick balancing. As summarized in an earlier section, this is consistent with

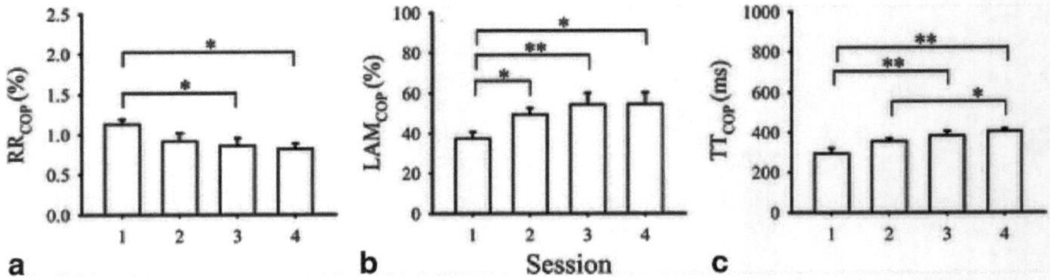

Fig. 8.3 Coupling strength and dynamical properties of the finger–COP (F–COP) interaction. **a** Recurrence rate (RR) of coupled F–COP trajectories by session. An increase in RR_{F-COP} reflects the tendency for collective control over finger and COP displacements. **b** LAM_{F-COP} measures transitions between coupled and uncoupled F–COP trajectories. Greater LAM_{F-COP} reflects an increase in the density of coupled F–COP trajectory segments. **c** Average length of coupled F–COP trajectories by session (TT_{F-COP}; unit: ms). *Error bars* are the within-subjects standard error of the mean (SEM). $^{*}P < 0.05$, $^{**}P < 0.01$, $^{***}P < 0.001$. (Reprinted with permission from Cluff et al. 2011 @ Springer)

previous work suggesting a facilitatory role for postural fluctuations (Balasubramaniam et al. 2000; Stoffregen et al. 1999, 2000). As seen in Fig. 8.3, we also demonstrated that the coupling strength between posture and hand displacement underwent substantial changes as a function of learning. Specifically, CRQA revealed that cross-recurrence, laminarity, and trapping time systematically changed with learning. Learning progressively stabilized the coupling between the upper limb and postural subsystems. In the following, we make the case that this progressive change in coupling emerged from the development of a hierarchical control system that can seamlessly switch between controlling the upper limb and postural systems (Newell et al. 2001).

According to the seminal motor learning model of Newell et al. (2001), individual subsystems become coupled to structural coordinative relationships to support performance. In this experiment, we showed the emergence of such coupling with learning. The improvement in performance can be attributed to two processes (1) at the level of individual subsystems (seen in learning sessions 1 and 2): decreases in regularity and discontinuity and (2) the lengthened coupling of the finger and COP displacements in the third learning session. Interpreting these results in the context of the model of Newell et al., one could argue that early learning changed the organization of individual subsystems and later learning influenced their coordinative relationship.

The increased coordination between the postural and finger subsystems could be due to either anticipatory (Flanagan and Wing 1997) or reactive mechanisms. We make the case that COP and finger displacement were controlled by a hierarchical system that was able to switch intermittently between individual and collective control of the subsystems in question. Future work should carefully examine how the individual biomechanical degrees of freedom of the hand and the postural system come together in a constrained way to achieve this functional coupling. We are presently engaged in research that would enable a comparison between end-effector analyses (such as the ones presented here) and the contribution of the many redundant degrees of freedom at the disposal of the CNS to the statistical properties of the end effectors during unstable object manipulation.

Attention and Task Performance

In the previous sections, I have looked at the changing landscape of control mechanisms that accompany motor learning in stick balancing. We have also seen in the preceding section about the nature of interactions between posture and hand displacements that reveal a hierarchical control structure. In recent years, several studies contributed to a generalized theory of attentional influences on motor performance (see, Wulf and Prinz 2001, for review). Other work has also examined the function of attention as being a deterrent to successful task performance (Beilock et al. 2002; Beilock et al. 2008; Beilock et al. 2008). Stemming from this research, the "constrained action" theory proposes that attention devoted to movement execution interrupts the automaticity of performance (Wulf et al. 2001). Performance, defined as the statistical stability or variability of motor execution, is dependent on whether attention is devoted to motor execution or outcome. As a test of the constrained action theory, Cluff et al. (2010), asked if specific task instructions would influence the way skilled practitioners carry out the stick-balancing task.

We employed six experimental conditions. (1) Posture baseline condition (P): quiet standing task for 30 s with no explicit instructions. (2) Posture-cognitive dual-task (P-CDT) condition: subjects performed upright standing while performing a silent, serial arithmetic task. (3) Posture-stick balancing (P-SB): subjects balanced the stick in upright stance without specific attentional instructions. (4) P-SB external focus condition (P-SBEXT): subjects were instructed to "minimize deviations of the stick from the vertical." (5) P-SB internal focus condition (P-SBINT): subjects were instructed to "focus on minimizing hand and finger movement." (6) P-SB-CDT: subjects performed a CDT while standing and balancing a stick. The methods used in this experiment are reported from Cluff et al. (2010).

We hypothesized that both postural and suprapostural components of the stick-balancing task would be stabilized by a task-irrelevant external focus of attention (P-SB-CDT). We predicted that an internal focus of attention would compromise dynamical stability in the stick-balancing task, resulting in variable COP and FINGER trajectories. In confirmation of the hypothesis, FINGER and COP trajectories were least variable when participants partitioned attentional resources between stick-balancing and cognitive task components, corresponding to an external, task-irrelevant focus (P-SB-CDT). In contrast, COP and FINGER displacements were least stable when the focus of attention was internal. Performance stability for the external, task-relevant condition was similar to control performance (P-SB). Figure 8.4 summarizes this effect.

It is interesting to note that our results did not directly support the constrained action theory proposed by Wulf et al. (2001). According to this theory, performance

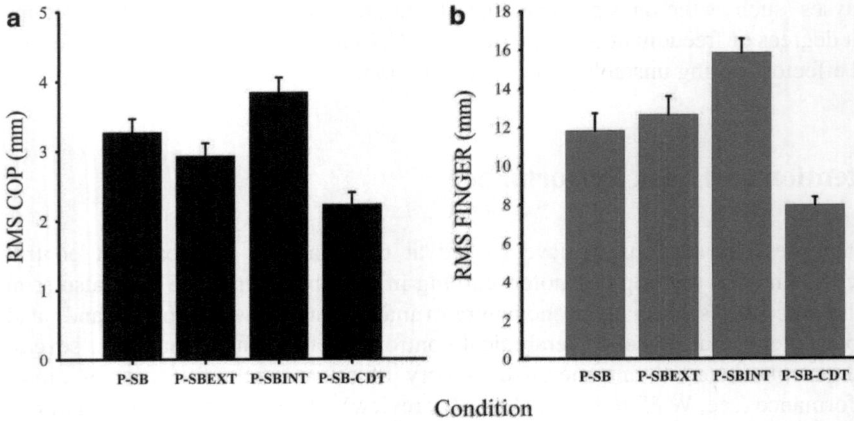

Fig. 8.4 The statistical stability of postural and suprapostural performance is dependent on focus of attention for balancing. COP and fingertip trajectories were least variable when stick balancing was performed with a concomitant cognitive load. **a** RMS COP and **b** RMS FINGER were reduced in the P-SB-CDT. Of particular interest was the stabilizing effect of cognitive load for stick-balancing performance. Finger trajectories were approximately half as variable in P-SB-CDT relative to other conditions. *Error bars* represent ± 1 SEM. (Reprinted with permission from Cluff et al. 2010 @ Springer)

variability is decreased, accompanied by increased frequency components when the attentional focus is external (minimizing movement of the stick). And conversely, performance variability is increased when attentional focus is internal (focus on minimizing finger displacements) with a slower frequency component dominating. Our data did not confirm these predictions. However, we showed that greatest reduction in performance variability was seen when performing a CDT, thus, taking the attentional focus away from the task of standing upright and concurrently balancing the stick.

It is important to underscore that previous studies of attentional focus did not deal with situations where there was an indistinct perceptual boundary between the body and the object being controlled. As originally noted by Gibson and later by investigators that study human and primate tool-use, handheld objects are often perceived as extensions of the body itself. This phenomenon, also known as exproprioception, needs to be considered in the context of the constrained action theory, proposed by Wulf and colleagues. Our results also showed that focusing on activity irrelevant to the physical task at hand (performing the concurrent cognitive task) was most beneficial to performance.

Note that all the participants in this study were skilled, having learned the stick-balancing task successfully. It is important to consider focus of attention in the context of Bernstein's ideas on expertise and its development. Although actors focus on moving body parts in the early stages of skill acquisition, attention shifts to wielded objects in the advanced stages of skill (Bernstein 1967). Advanced tennis players tend to focus on the ball or end point of the trajectory for a successful return, rather than the racquet or limb. In the stick-balancing case, there is no clear boundary between where one ends and the other begins. Therefore, it is likely that stick balancers at earlier stages of skill acquisition show stronger differences as a function of attentional focus. Following this study, one could predict that task-irrelevant focus would not benefit less experienced stick balancers (Milton et al. 2008).

General Discussion

Goal-directed motor tasks commonly require the use of objects, tools, and implements to interact with our environment. Dynamic object interactions can vary in terms of the rigidity, geometry, and stability of manipulated objects, yet, we formulate adaptive motor responses that accommodate differences in the task, context, and object mechanics. Knowledge of the underlying control mechanisms and learning processes is imperative for understanding the basis of skilled object manipulation. The four studies presented in this review chapter used an inverted pendulum (i.e., stick) balancing paradigm to investigate skill acquisition and elaborate the task- and context-dependent attributes of unstable object control.

In the first study, we evaluated the statistics of the spatio-temporal properties of stick displacement (Cluff and Balasubramaniam 2009). After establishing that learning resulted in a systematic increase of balancing time, we fit our data to theoretical Lévy distributions. Results showed the probability of fingertip speed increase over analyzed time scales was Lévy-distributed and that this distribution changed with learning. Essentially, motor learning caused systematic increase in the prevalence of upper limb displacements in the standing condition, a feature less visible in seated subjects.

Motivated in part by the observations of Cabrera and Milton (2002) that angular stick fluctuations occur on timescales shorter than estimated voluntary control delays ($\sim$100 ms) and show amplitude variations that are characteristic of on–off intermittency, we quantified the nature of this intermittency using modern analytic tools based on a numerical phase space reconstruction method (RQA). Provided that angular stick fluctuations are intermittent, we hypothesized that upper limb displacements would be composed of two independent timescale components differentiated by their correlative properties. We additionally hypothesized that the temporal structure of upper limb corrections would be modulated by the balancing context. We used a numerical phase space reconstruction method (cf. Webber and Zbilut 1992; Marwan et al. 2007) to determine whether the switching time to feedback control was dependent on motor learning and the balancing context. We demonstrated (Cluff et al. 2009) that upper limb displacements are indeed composed of two independent timescale components: a fast stochastic component and slow closed-loop feedback control. Our results revealed that the discontinuity, stability, and regularity of upper limb displacements changed systematically across training sessions. Another important finding was the differential control evoked by changes in the balancing context. We found that the average time interval between upper limb corrections was substantially shorter for the seated balance.

Our studies raise important concerns for studies that employ the Kalman filter algorithm (Kalman 1960), commonly seen in studying systems with assumed motor and sensory noise. The limitation of the Kalman filter is that it is designed to handle Gaussian-distributed additive noise. I believe that the discovery of Lévy distributions in fingertip displacements questions the validity of using control models used by Mehta and Schaal (2002). I strongly suggest the use of suitable estimation algorithms (Gordon et al. 2006; Sinha et al. 2007) that are designed to be sensitive to Lévy distributions. Although it is well established that variability in muscle force production increases with movement amplitude, state-dependent motor noise is commonly assumed to be negligible (for review, see Todorov 2005). Future studies should carefully consider the role of stochasticity in the CNS in the development of state estimation models.

Complex motor tasks often involve the coordination of posture and voluntary arm movements. In the third study, we performed an innovative analysis that investigated the learning-dependent coupling of posture and upper limb dynamics (cf. Marwan et al. 2007). We interpreted our results from the perspective of a hierarchical learning model (Newell et al. 2001) and this study was among the first to quantify learning and control at multiple levels of the motor system. Our results corroborated the model of Newell et al. (2001) and demonstrated that skill acquisition involved two independent learning processes. First, we found that posture and upper limb control were governed by intermittent balancing strategies and that the time interval between corrections increased systematically across the investigated training period. The second learning effect involved the incremental occurrence and lengthened the coupling of correlated posture-upper limb trajectories.

Thus, I can make the case in which posture-upper limb coordination is contingent on a state-dependent (de)coupling mechanism. I have additionally proposed that the abrupt decoupling of posture-upper limb trajectories is caused by the instability of subsystem dynamics. Our rationale is supported by the observation that human feedback parameters are often tuned near instability. Thus, we anticipate that perturbing forces applied to the tip of the stick would induce the abrupt dissociation of posture-upper limb coupling to enable independent subsystem corrections. State-dependent posture-upper limb coupling is a plausible control mechanism and is similar to the spontaneous recruitment of body segments that has been shown to intermittently stabilize intereffector coordination (Buchanan and Kelso 1999).

Finally, I tested the constrained action theory (Wulf and Prinz 2001) using explicit attentional manipulations during unstable object control. Performance variability increased irrespective of whether participants focused on minimizing fingertip or stick displacements. Thus, explicit instruction (relevant to the task at hand), increased task variability and was in fact counterproductive (Beilock et al. 2002, 2008; Beilock et al. 2008). However, the performance of a task-irrelevant cognitive task reduced performance variability of both stick-balancing and COP displacements. The critical time for switching to corrective movements also increased for both outcome and execution-oriented attentional manipulations. It is likely that participants shifted to a slower and possibly more conscious corrective mechanism. Note that the general failure to support the constrained action theory might have been due to the fact that our participants were expert stick balancers. Novice subjects are more likely to be influenced by this manipulation (Beilock et al. 2002) since they lack specific knowledge about the interaction between limb and object dynamics. Such participants might gain more from outcome-oriented feedback (Todorov et al. 1997; Malone and Bastian 2010).

The stick-balancing problem is thus a rich problem that provides the opportunity to explore a variety of issues in motor control and learning. By employing techniques that have borrowed from statistical mechanics, we have established how task-specific changes can be observed at different spatial and temporal scales as a person learns to master this complex task. Future research should reveal the role of multiple joints and the collective error correction formed to solve the stick's dynamics and experimentation that involves manipulating the physical dynamics of the stick itself. This research will further add to our understanding of the complex interactions between the task, actor, and environment during the acquisition and performance of unstable object manipulation.

References

Ahmed AA, Wolpert DM (2009) Transfer of dynamic learning across postures. J Neurophysiol 102(5):2816–2824

Ahmed AA, Wolpert DM, Flanagan JR (2008) Flexible representations of dynamics are used in object manipulation. Curr Biol 18(10):763–768

Balasubramaniam R, Riley MA, Turvey MT (2000) Specificity of postural sway to the demands of a precision task. Gait Posture 11(1):12–24

Balasubramaniam R, Turvey MT (2004) Coordination modes in the multisegmental dynamics of hula-hooping. Biol Cybern 90(3):176–190

Balasubramaniam R, Wing AM (2002) The Dynamics of Standing Balance. Trends Cogn Sci 6(12):531–536

Beilock SL, Bertenthal BI, Hoerger M, Carr TH (2008) When does haste make waste? Speed-accuracy tradeoff, skill level and the tools of the trade. J Exp Psychol Appl 14(4):340–352

Beilock SL, Carr TH, MacMahon C, Starkes JL (2002) When paying attention becomes counterproductive: impact of divided versus skill-focused attention on novice and experienced performance of sensorimotor skills. J Exp Psychol Appl 8:6–16

Bernstein NA (1967) The coordination and regulation of movements. Oxford, Pergamon

Bingham GP (1988) Task-specific devices and the perceptual bottleneck. Hum Mov Sci 7:225–264

Boulet J, Balasubramaniam R, Daffertshofer A, Longtin A (2010) Stochastic two-delay differential model of delayed visual feedback effects on postural dynamics. Philos Trans R Math Phys Eng Sci 368(1911):423–438

Buchanan JJ, Kelso JAS (1999) To switch or not to switch: recruitment of degrees of freedom stabilizes biological coordination. J Mot Behav 31(2):126–144

Cabrera JL, Luciani C, Milton J (2006) Neural control on multiple time scales: insights from human stick balancing. Condens Matter Phys 9(2):373–383

Cabrera JL, Milton JG (2002) On-off intermittency in a human balancing task. Phys Rev Lett 89(15):158702

Cabrera JL, Milton JG (2004a) Stick balancing: on-off intermittency and survival times. Nonlinear Stud 11:305–317

Cabrera JL, Milton JG (2004b) Human stick balancing: tuning Lévy flights to improve balance control. Chaos 14(3):691–698

Cluff T, Robertson DGE, Balasubramaniam R (2008) Kinetics of hula hooping: an inverse dynamics analysis. Hum Mov Sci 27(4):622–635
Cluff T, Balasubramaniam R (2009) Motor learning characterized by changing Lévy distributions. PLoS ONE 4:e5998
Cluff T, Riley MA, Balasubramaniam R (2009) Dynamical structure of hand trajectories during pole balancing. Neurosci Lett 464(2):88–92
Cluff T, Gharib T, Balasubramaniam R (2010) Attentional influences on the performance of secondary tasks during postural control. Exp Brain Res 203(4):647–658
Cluff T, Boulet JW, Balasubramaniam R (2011) Task-specific coupling between posture and hand displacements develop while learning a novel visuomotor balancing task. Exp Brain Res 213: 55–65
Collins JJ, De Luca CJ (1994) Random walking during quiet standing. Phys Rev Lett 73(5):746–768
Collins JJ, De Luca CJ (1995) Upright, correlated random walks: a statistical-biomechanics approach to the human postural control system. Chaos 5(1):57–63
Diedrichsen J (2007) Optimal task-dependent changes of bimanual feedback control and adaptation. Curr Biol 17(19):1675–1679
Faisal AA, Selen LPJ, Wolpert DM (2008) Noise in the nervous system. Nat Rev Neurosci 9(4):292–303
Foo P, Kelso JAS, de Guzman GC (2000) Functional stabilization of unstable fixed points: human pole balancing using time to balance information. J Exp Psychol Hum Percept Perform 26(4):1281–1297
Gawthrop P, Loram I, Lakie M, Gollee H (2011) Intermittent control: a computational theory of human motor control. Biol Cybern 104(1–2):31–51
Gordon N, Percival J, Robinson M (2006) The Kalman-Levy filter and heavy-tailed models for tracking maneuvering targets. In: Proceedings of the 6th International Conference on Information Fusion
Hollerbach JM, Flash T (1982) Dynamic interactions between limb segments during planar arm movements. Biol Cybern 44(1):67–77
Imamizu H, Kuroda T, Miyauchi S, Yoshioka T, Kawato M (2003) Modular organization of internal models in the human cerebellum. Proc Natl Acad Sci USA 100(9):5461–5466
Jeka JJ, Oie KS, Kiemel T (2000) Multisensory information for human postural control: integrating touch and vision. Exp Brain Res 134(1):107–125
Kalman RE (1960) A new approach to linear filtering and prediction problems. Trans ASME J Basic Eng 82:35–45
Kiemel T, Oie KS, Jeka JJ (2002) Multisensory fusion and the stochastic structure of postural sway. Biol Cybern 87(4):262–277
Kurtzer I, DiZio P, Lackner J (2003) Task-dependent motor learning. Exp Brain Res 153(1):128–132
Kurtzer IL, Pruszynski JA, Scott SH (2008) Long-latency reflexes of the human arm reflect an internal model of limb dynamics. Curr Biol 18(6):449–453
Lackner JR, DiZio P (1994) Rapid adaptation to Coriolis force perturbations of arm trajectory. J Neurophysiol 72(1):299–313
Loram ID, Lakie M (2002) Human balancing of an inverted pendulum: position control by small, ballistic-like, throw and catch movements. J Physiol 540:1111–1124
Loram ID, Gawthrop PJ, Lakie M (2006) The frequency of human, manual adjustments in balancing an inverted pendulum is constrained by intrinsic physiological factors. J Physiol 577:417–432
Loram ID, Lakie M, Gawthrop P (2009) Visual control of stable and unstable loads: what is the feedback delay and extent of linear time-invariant control? J Physiol 587:1343–1365
Loram ID, Gollee H, Lakie M, Gawthrop PJ (2011) Human control of an inverted pendulum: is continuous control necessary? Is intermittent control effective? Is intermittent control physiological? J Physiol 589:307–324
Mah CD, Mussa-Ivaldi FA (2003a) Evidence for a specific internal representation of motion-force relationships during object manipulation. Biol Cybern 88(1):60–72
Mah CD, Mussa-Ivaldi FA (2003b) Generalization of object manipulation skills learned without limb motion. J Neurosci 23(12):4821–4825

Malone LA, Bastian AJ (2010) Thinking about walking: effects of conscious correction versus distraction on locomotor adaptation. J Neurophysiol 103(4):1954–1962

Marin L, Bardy BG, Baumberger B, Fluckiger M, Stoffregen TA (1999) Interaction between task demands and surface properties in the control of goal oriented stance. Hum Mov Sci 18(1):31–47

Marwan N, Carmen Romano M, Thiel M, Kurths J (2007) Recurrence plots for the analysis of complex systems. Phys Rep 438(5–6):237–329

Mehta B, Schaal S (2002) Forward models in visuomotor control. J Neurophysiol 88(2):942–953

McNevin NH, Shea CH, Wulf G (2003) Increasing the distance of an external focus of attention enhances learning. Psychol Res 67(1):22–29

Mehta B, Schaal S (2002) Forward models in visuomotor control. J Neurophysiol 88(2):942–953

Milner TE, Franklin DW, Imamizu H, Kawato M (2006) Central representation of dynamics when manipulating handheld objects. J Neurophysiol 95(2):893–901

Milton JG, Ohira T, Cabrera JL, Fraiser RM, Gyorffy JB, Ruiz FK, Strauss MA, Balch EC, Marin PJ, Alexander JL (2009) Balancing with vibration: a preclude for drift and act balance control. PLoS ONE 4(10):e7427

Milton JG, Small SS, Solodkin A (2004) On the road to automatic: dynamic aspects in the development of expertise. J Clin Neurophysiol 21(3):134–143

Milton JG, Small SS, Solodkin A (2008) Imaging motor imagery: methodological issues related to expertise. Methods 45(4):336–341

Narendra KS, Annaswamy, AM (1986) Robust Adaptive Control in the Presence of Bounded Disturbances. IEEE Trans Autom Control 31(4):306–315

Newell KM, Liu YT, Mayer-Kress G (2001) Time scales in motor learning and development. Psychol Rev 108:57–82

Peterka RJ, Loughlin PJ (2004) Dynamic regulation of sensorimotor integration in human postural control. J Neurophysiol 91(1):410–423

Riccio GE, Stoffregen TA (1988) Affordances as constraints on the control of stance. Hum Mov Sci 7(2–4):265–300

Riley MA, Stoffregen TA, Grocki MJ, Turvey MT (1999) Postural stabilization for the control of touching. Hum Mov Sci 18(6):795–817

Seidler RD, Bloomberg JJ, Stelmach GE (2001) Patterns of transfer of adaptation among body segments. Behav Brain Res 122(2):145–157

Shadmehr R, Mussa-Ivaldi FA (1994) Adaptive representation of dynamics during learning of a motor task. J Neurosci 14(5):3208–3224

Sinha A, Kirubarajan T, Bar-Shalom Y (2007) Application of the Kalman-Levy filter for tracking maneuvering targets. IEEE Trans Aerosp Electron Syst 47:1099–1147

Singh K, Scott SH (2003) A motor learning strategy reflects neural circuitry for limb control. Nat Neurosci 6(4):399–403

Stoffregen TA, Pagulayan RJ, Bardy BG, Hettinger LJ (2000) Modulating postural control to facilitate visual performance. Hum Mov Sci 19(2):203–220

Stoffregen TA, Smart LJ, Bardy BG, Pagulayan RJ (1999) Postural stabilization of looking. J Exp Psychol Hum Percept Perform 25:1641–1658

Ting LH, Macpherson JM (2005) A limited set of muscle synergies for force control during a postural task. J Neurophysiol 93(1):609–613

Todorov E (2005) Stochastic optimal control and estimation methods adapted to the noise characteristics of the sensorimotor system. Neur Comput 17(5):1084–1108

Todorov E, Jordan M (2002) Optimal feedback control as a theory of motor coordination. Nat Neurosci 5(11):1226–1235

Todorov E, Shadmehr R, Bizzi E (1997) Augmented feedback presented in a virtual environment accelerates learning of a difficult motor task. J Mot Behav 29(2):147–158

Treffner PJ, Kelso JAS (1999) Dynamic encounters: long-term memory during functional stabilization. Ecol Psychol 11(2):103–137

Vereijken B, van Emmerik REA, Bongaardt R, Beek WJ, Newell KM (1997) Changing coordinative structures in complex skill acquisition. Hum Mov Sci 16(6):823–844

Vereijken B, van Emmerik REA, Whiting HTA, Newell KM (1992) Free(z)ing degrees of freedom in skill acquisition. J Mot Behav 24:133–142
Webber CL Jr, Zbilut JP (1992) Embeddings and delays as derived from quantification of recurrence plots. Phys Lett A 171(3–4):199–203
Wolpert DM, Miall RC, Winter JL, Stein JF (1992) Evidence for an error deadzone in compensatory tracking. J Mot Behav 24(4):299–308
Wolpert DM, Ghahramani Z, Jordan MI (1995) An internal model for sensorimotor integration. Science 269(5232):1880–1882
Wolpert DM, Miall RC, Kawato M (1998) Internal models in the cerebellum. Trends Cogn Sci 2(9):338–347
Wulf G, McNevin NH, Shea CH (2001) The automaticity of complex motor skill learning as a function of attentional focus. Q J Exp Psychol A 54(4):1143–1154
Wulf G, Prinz W (2001) Directing attention to movement effects enhances learning: a review. Psychon Bull Rev 8(4):648–660
Zanone PG, Kelso JAS (1992) The evolution of behavioral attractors with learning: nonequilibrium phase transitions. J Exp Psychol Hum Percept Perform 18(2):403–421
Zatsiorsky VM, Duarte M (1999) Instant equilibrium point and its migration in standing tasks: rambling and trembling components of the stabilogram. Motor Control 3(1):28–38

Chapter 9
Intermittent Motor Control: The "drift-and-act" Hypothesis

John G. Milton

Introduction

The voluntary movements of a skilled individual, such as an athlete, dancer, martial artist or musician, are a marvel to behold: precise and reproducible while at the same time smooth and effortless. Yet this impression of perfect motor control does not simply reflect the output of a "better wired" nervous system. Practice is required to both acquire and maintain the skill level. Moreover, under appropriately stressful conditions, skilled performances can rapidly deteriorate, a phenomenon known as *choking* (Beilock 2011). These observations strongly suggest that expert motor control must, at least in part, have a dynamical basis. Surprisingly, all of this is accomplished by a nervous system that is both noisy and delayed (Milton 2011). How is this possible?

Three observations motivate a reassessment of the role of sensory feedback in motor control: (1) the performance of movements is enhanced by sensory feedback (Kuiken et al. 2007; Suminski et al. 2010); (2) time-delayed feedback is capable of providing fast (Cabrera and Milton 2002), and even anticipatory (Voss 2000; Stepp 2009), control; and (3) noise has beneficial effects on motor control (Priplata et al. 2002; Matthews 1996; Harris and Wolpert 1998). The fundamental difficulty for controlling a complex voluntary movement using a time-delayed and noisy nervous system is deciding whether an observed departure from a goal should be acted upon or not (Milton et al. 2008). One strategy is to wait before making a response, thereby relying on the possibility that the deviation might be counteracted just by chance. On the other hand, there is the temptation to act quickly and correct all deviations. Here the problem is that a deviation observed at time t cannot be corrected until time $t+\tau$ later, where τ is a delay. Thus it becomes possible that the corrective movement and the noisy perturbation at time $t+\tau$ are in the same direction. This leads to the phenomenon of "over control" which is typically destabilizing. Control theoretic arguments suggest that the best control strategy in the presence of noise

---

J. G. Milton (✉)
W. M. Keck Science Department, The Claremont Colleges, Claremont, CA 91711, USA
e-mail: jmilton@jsd.claremont.edu

M. J. Richardson et al. (eds.), *Progress in Motor Control,*
Advances in Experimental Medicine and Biology 782,
DOI 10.1007/978-1-4614-5465-6_9,

is an intermittent one in which active corrections are made only when deviations exceed a threshold. This conclusion is true whether the time delay is zero (Todorov and Jordan 2002; Todorov 2004) or not zero (Stépán and Insperger 2006; Insperger 2007; Bottaro et al. 2008; Milton et al. 2008; Asai et al. 2009; Milton et al. 2009a, 2009c; Gawthrop et al. 2011).

An experimental paradigm for investigations into the role played by noise and delay in the control of a complex voluntary task is stick balancing in 1-D (Foo et al. 2000; Jirsa et al. 2000), 2-D (Mehta and Schaal 2002; Bormann et al. 2004; Cabrera et al. 2004; Patzelt and Pawelzik 2011; Milton et al. 2011), and 3-D (Cabrera and Milton 2002; Mehta and Schaal 2002; Cabrera and Milton 2004a, 2004b; Cluff and Balasubramaniam 2009; Milton et al. 2009a, 2009b; Cluff et al. 2010, 2011). Stick balancing is particularly well suited for studying the development of skill and expertise since skill levels can be increased 10–100 fold within a few days of practice. The first section takes advantage of graphical techniques to illustrate basic concepts of time-delayed feedback control. It is argued that the effects of noise are anticipated to be most important when control is tuned near the "edge of stability". The next section introduces the basic paradox of the development of stick-balancing skill, namely, even under conditions where negative feedback is ineffective, skill increases dramatically with practice. The next two sections argue that the interplay between noise and delay can make stick balancing easier to perform. This possibility potentially minimizes reliance on anticipatory, or predictive, control strategies. First, near the edge of stability, the noise can postpone the onset of instability and also can result in a passive type of control in which corrective movements are made intermittently. Second, the *drift and act hypothesis* extends these ideas to situations in which control is achieved using a "*safety net*" network topology. In this context, corrective movements are made intermittently as thresholds are crossed. The final section directs attention to a number of outstanding issues.

Background

All of the movements made by an organism must ultimately be consistent with Newton's laws of classical mechanics. These equations of motion account not only for forces generated by muscle activations, but also for those forces related to the biomechanical properties of the body and those related to the environment in which the movements occur (Chiel and Beer 1997; Milton et al. 2004). Consequently, mathematical models for the feedback control of movement are differential equations of the general form

$$\ddot{x}(t) + b\dot{x}(t) + cx(t) = F_{control}(x(t), \dot{x}(t), \ddot{x}(t)) \tag{9.1}$$

where b, c are positive constants and $x(t), \dot{x}(t), \ddot{x}(t)$ are, respectively, the displacement, velocity, and acceleration of the controlled variable. The function $F_{control}$ describes the action of the proposed controller. I have introduced the notation $\dot{x}(t) \equiv dx/dt$ and $\ddot{x}(t) \equiv d^2x/dt^2$.

Neural control of movement adds two complexities to $F_{control}$. First, neural controllers are time-delayed. The time delays arise because axonal conduction velocities are finite and distances between neurons and muscle cells can be appreciable. Reaction times, a basic measure of neural processing speed (Gabbard 2004), can be of the order of hundreds of milliseconds and increase further as task complexity increases (Hick 1952).

Second, the nervous system behaves as a very noisy environment (Areili et al. 1996; Fatt and Katz 1950; Riani and Simonotto 1994). The sources of noise range from the fluctuations in the opening and closing of ion channels to "chaotic noise" generated by neural circuits such as the recurrent inhibitory loop (Mackey and an der Heiden 1984). Moreover noise can enter as an approximation to the large number of uncorrelated inputs ($\approx 10^4$) that a neuron in the cortex receives (Eq. 9.1). The effects of noise on motor control cannot be neglected since noise has been shown to have beneficial effects (Priplata et al. 2002; Matthews 1996; Harris and Wolpert 1998).

Taken together, these observations suggest that Eq. 9.1 becomes the delay differential equation (DDE)

$$\ddot{x}(t) + b\dot{x}(t) + ax(t) = F_{control}(x(t-\tau), \dot{x}(t-\tau), \ddot{x}(t-\tau), \xi(t)) \qquad (9.2)$$

where $\xi(t)$ describes the noisy inputs and for simplicity we have replaced a measured distribution of neural delays with the mean value, τ. In the absence of noise, feedback depends on both x and $\dot{x}$, namely we have so-called PD-control (Stepan 2009). However, the measurement of $\dot{x}$ in the presence of noise is problematic and thus the role of $\dot{x}$ in neural control remains uncertain. Recently, it has been argued that the feedback for balance control also depends on $\ddot{x}$ (Lockhart and Ting 2007). Indeed neural mechanoreceptors measure force. However, a consideration of the effects of noise on the properties of neutral DDEs, i.e., differential equations in which the delay occurs in the highest derivative, is beyond the scope of this discussion.

The effects of delay on control can be illustrated by considering the first-order delay differential equation (DDE) with negative feedback (Fig. 9.1a). Well studied examples include, for example, the pupil light reflex (Longtin and Milton 1989; Bressloff et al. 1996; Bressloff and Wood 1997), respiration (Mackey and Glass 1977), gene expression (Yildirim et al. 2004), and blood cell dynamics (Mackey 1979). The DDE takes the form

$$\dot{x}(t) + kx(t) = F(x(t-\tau)) \qquad (9.3)$$

where k is a positive constant, $x(t)$, $x(t-\tau)$ are, respectively, the values of the state variable at times t, $t-\tau$ and F describes negative feedback, namely feedback that acts to decrease its own value. In order to obtain a unique solution to Eq. 9.3, it is necessary to choose an appropriate initial condition. When $\tau = 0$ the initial condition corresponds to the value of $x(t)$ measured at a single instance in time, $x(t_0)$. However, when $\tau > 0$, the initial condition takes the form of an initial function defined over a time interval of length τ, namely, $\Phi(s), s \in [-\tau, 0]$. If we define the dimension of a dynamical system as the number of initial values that must be specified to uniquely

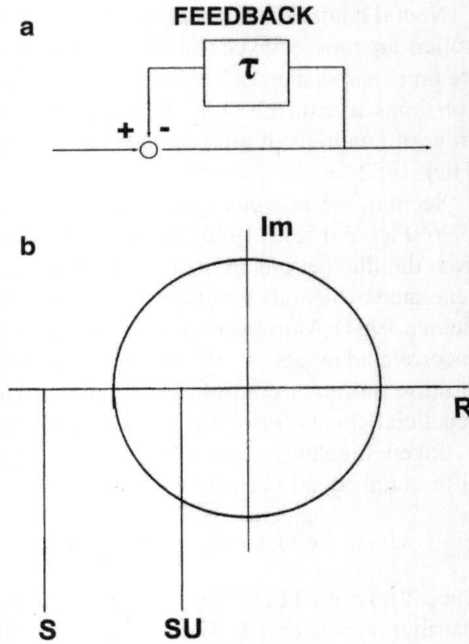

Fig. 9.1 **a** Schematic representation of a time-delayed negative feedback control mechanism. **b** Graphical representation of the stability of Eq. 9.3 in the complex plane. *S* stable for all delays, *Su* instability occurs when there is an intersection

determine the solution, then we see that Eq. 9.3 corresponds to an infinite dimensional differential equation.

The fixed-point, or control set point, of Eq. 9.3, x^*, is obtained by setting $\dot{x}(t) = 0$ and then solving the equation

$$kx^* = F(x^*)$$

In order to determine the stability of the fixed-point, namely the response of Eq. 9.3 following a perturbation away from x^*, we define $u(t) = x(t) - x^*$ and then linearize Eq. 9.3 about the fixed-point to obtain

$$\dot{u}(t) + ku(t) = du(t - \tau) \tag{9.4}$$

where the slope of f evaluated at x^* is $d < 0$ since we have negative feedback.

Next we assume that solutions of Eq. 9.4 are of the form $u(t) \approx e^{\lambda t}$. Substituting into Eq. 9.4, we obtain the characteristic equation

$$\lambda + k = de^{-\lambda\tau} \tag{9.5}$$

In general, λ is complex, namely $\lambda = \gamma + jf$ where $j = \sqrt{-1}$ and f is the frequency. Hence the transcendental equation (Eq. 9.5) has an infinite number of solutions.[1]

[1] The transcendental nature of (Eq. 9.5) can be readily seen by writing $e^{-2\pi j\omega\tau} = \cos(2\pi\omega\tau) - j\sin(2\pi\omega\tau)$.

The condition for stability is that the real part, γ, of all of the eigenvalues (real and complex) must be negative. The stability boundary corresponds to $\gamma = 0$. Using this observation, we can re-write Eq. 9.5 as

$$-\frac{k}{d} + j\frac{f}{d} = e^{-jf\tau} \tag{9.6}$$

As is shown in Fig. 9.1b, this equation can be solved graphically in the complex plane: the left-hand side is a straight line and the right-hand side is a circle (MacDonald 1989). When the straight line intersects the circle, the fixed-point, x^*, loses stability and a stable limit cycle appears for physiologically relevant choices of F (Longtin and Milton 1989). Clearly whether instability occurs depends on the interplay between τ and the reflex gain

$$G \equiv \left(\frac{k}{d}\right)^{-1}$$

If $G < 1$, i.e., $k > d$, x^* is stable for all τ. When $G > 1$ the fixed-point can lose stability.

A sudden change in stability as a parameter (in this case G or τ) is changed is called a *bifurcation*. Bifurcations, in turn, are classified in terms of the behaviors that arise when the fixed-point becomes unstable. In the case of Eq. 9.3, there is an exchange of stability: the fixed-point becomes unstable and a stable limit appears (Longtin and Milton 1989). Bifurcations of this type in which a limit cycle is created are called Hopf bifurcations: in this case it is a supercritical Hopf bifurcation (see the following section).

The prediction that an oscillation arises when the fixed-point of a time-delayed negative feedback control loop becomes unstable can be confirmed experimentally in the pupil light reflex by focusing a narrow beam of light onto the edge of the pupil (Stark and Cornsweet 1958). This corresponds to high-gain feedback since light either reaches the retina or is blocked by the iris. The period, T, of these oscillations, referred to as the pupil cycle time, is $\sim$900–1000 ms and is in agreement with the prediction of Eq. 9.3 that $2\tau < T < 4\tau$ (Longtin and Milton 1989).

Our focus is on the interplay between random perturbations ("noise") and τ on control. There are two distinct ways in which noise can influence dynamics. Additive noise refers to the situation that the effects of noise are independent of the state of the system and hence Eq. 9.3 becomes

$$\dot{x}(t) + kx(t) = f(x(t-\tau)) + \xi(t)_A \tag{9.7}$$

where $\xi(t)_A$ is, for example, Gaussian distributed white noise. The alternative is that noise affects the system in a state-dependent, or multiplicative, manner for which Eq. 9.3 becomes

$$\dot{x}(t) + kx(t) = f(x(t-\tau)), \xi(t)_M) \tag{9.8}$$

A well-known example of multiplicative noise arises in the Hodgkin-Huxley model of a neuron. Membrane noise reflects fluctuation in conductance: its effects on current are proportional to the product of conductance and membrane potential and hence are state-dependent. Equations 9.7 and 9.8 are stochastic delay differential equations and the effects of noise must be taken into account to obtain a solution. In contrast, the uncertainties in measurements of a variable are added to the solution of the dynamical system of equations. Although the effects of multiplicative and additive noise differ when $\tau = 0$, they are often quite similar when $\tau \neq 0$ (Longtin et al. 1990; for a possible exception see Sect. 4.2). This observation may be related to the equivalence of the effects of additive and multiplicative noise for discrete dynamical systems (Crutchfield et al. 1982; Linz and Lücke 1986): a DDE reduces to a discrete map in the singular perturbation limit.

A mantra of modern day dynamical systems theory is that complex dynamical systems tend to self-organize near stability boundaries (Guckenheimer and Holmes 1983; Venkadesan et al. 2007). This scenario is anticipated to emphasize effects that arise from the interplay between noise and delay. Let $\tau = 0$ and consider the response to a perturbation as a control parameter, μ, approaches the critical value at which the supercritical Hopf "bifurcation" occurs, μ_0, from below: perturbations away from the stable fixed-point are damped out less and less quickly. This "slowing down" phenomenon is a manifestation of the fact that the real part of the principal eigenvalue approaches 0 as $\mu \to \mu_0$. At μ_0, noise dominates the dynamics since the stable limit cycle has zero amplitude! However, it is also true that a system tuned at, or near to, μ_0 can change very quickly in response to a perturbation (Freeman et al. 2006). Indeed the control of high performance jet fighters is tuned in this manner for exactly this purpose. Even in the case where $\tau \neq 0$, the dynamics are dominated by the critical behaviors that arise as the real part of the first complex eigenvalue pair to become unstable becomes zero (Longtin et al. 1990; Ushida 2011). Thus the dynamics of a negative feedback control loop tuned near the edge of stability can primarily reflect the interplay between noise, delay, and possibly certain nonlinearities in the dynamic system. Experimental and computational observations indicate that this interplay between noise and delay can itself stabilize, at least transiently, the unstable fixed-point (Cabrera and Milton 2002; Milton et al. 2008, 2009c).

The Control Dilemma

The challenge faced by a stick balancer is to maintain a stick balanced in the upright position by moving their fingertip. The problem is that the upright position is unstable: even the slightest deviation from vertical causes the pendulum to topple over. The ability of a control mechanism to stabilize an inverted pendulum is a widely used benchmark: the better the control, the more robust the controller. Thus the mathematical observation that the upright position can be stabilized using time-delayed negative feedback has fueled an extensive literature on this subject (for recent reviews see Stepan 2009 and Milton et al. 2009b). The present day "state of the art" mathematical models assume that (1) the feedback is linear (for notable exceptions see Sieber and Krauskopf 2004a, b); (2) the feedback functions as a PD-controller (Stepan 2009); and (3) the upright position corresponds to a stable fixed-point of the

Fig. 9.2 Comparison of the neural delay, τ_n, to the critical delay calculated using Eq. 9.9 (*solid line*). The methods used to estimate τ_n were response to mechanism perturbation (□), cross-correlation (o) and response to a visual blank-out (Δ)

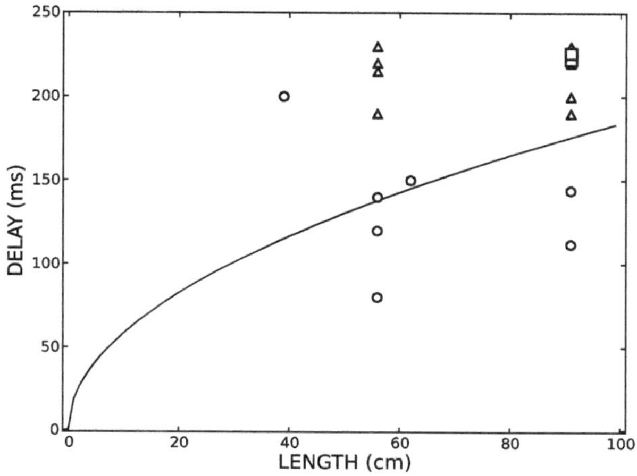

dynamical system. Irrespective of the details of the model, an essential condition for stability is that the neural feedback delay, τ_n, be less than a critical delay, τ_c, that depends on the length, ℓ, of the stick (Stepan 2009; Milton et al. 2009b)

$$\tau_n < \tau_c = \sqrt{\frac{\ell}{3g}} \tag{9.9}$$

where g is the acceleration due to gravity.

Time Delays for Stick Balancing

Three approaches have been tried to estimate the time delay for stick balancing: (1) measurement of the response time to a small mechanical perturbation (Mehta and Schaal 2002); (2) measurement of the recovery time following a prolonged visual blank out (R. Meyer and J. Milton in preparation); and (3) calculation of the cross-correlation between the position of the stick and the corrective movement made by the fingertip (Cabrera and Milton 2004a; Milton et al. 2009a). Figure 9.2 compares the measured τ_n versus τ_c as a function of ℓ. As can be seen the estimates of τ_n are typically longer than τ_c. The exceptions occur when τ_n is measured from the cross-correlation function. Estimates of the delay using the cross-correlation function are currently considered to be less reliable than these other approaches (van der Kooij et al. 2005): for a different opinion see (Berger et al. 1989). Thus in the following discussion, we do not consider latencies measured using the cross-correlation function.

The stick lengths commonly studied for stick balancing range from 0.29–0.91 m, most often ~ 0.55–0.56 m in our laboratory. These stick lengths are too short to permit the existence of a stable upright position using negative feedback. If we assume that $\tau_n \sim 225$ ms, then stability requires that $\ell \geq 1.49$ m. However, it should be noted

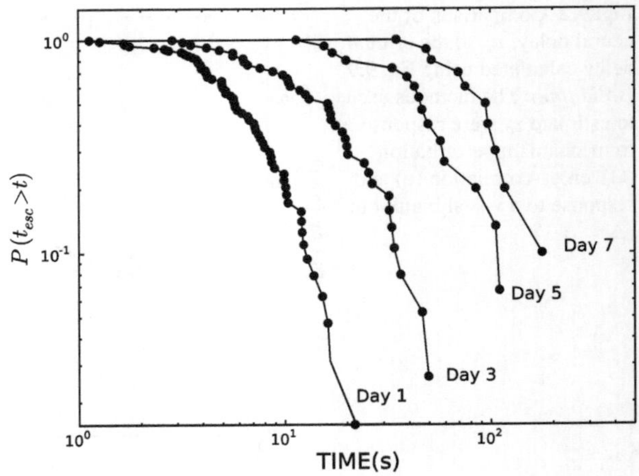

Fig. 9.3 Stick balancing survival curves at the fingertip as a function of days of practice for one subject using a 0.56 m stick

that the stability criterion given by Eq. 9.9 is derived for stick balancing in 1-D. Symmetry considerations suggest that this condition would be similar in 2-D (T. Insperger, personal communication) and perhaps would be slightly shorter in 3-D (S. A. Campbell, personal communication). In the remainder of the discussion, we assume that $\tau_n > \tau_c$.

Survival Functions

The dominant feature of stick balancing, especially for novices, is that the stick eventually falls. A key concept for the analysis of an unstable and noisy dynamical system is the *first passage time* (Milton et al. 2008). The first passage time is the time that it takes for a trajectory starting close to the unstable fixed-point to exceed a threshold located a finite distance from the unstable fixed-point. The *survival function* provides an estimate of the probability that a given stick balancing trial (a realization of a first passage time) lasts longer than time t.

The stick survival function can be estimated by, for example, measuring the time it takes for the stick to fall for, let us say, 25 consecutive stick balancing trials. The survival function is a plot of the fraction of sticks still balanced at time t, $P(t_{esc} > t)$, as a function of time (Fig. 9.3). These survival curves resemble the Weibull survival function

$$P(t_{esc} > t) \approx \exp[-(kt)^{\alpha}]$$

where α and k are positive constants (Cabrera and Milton 2004b; Cabrera et al. 2006; Cabrera and Milton 2012). These observations emphasize that stick balancing at the fingertip is best described as a transient stabilization of an unstable upright position.

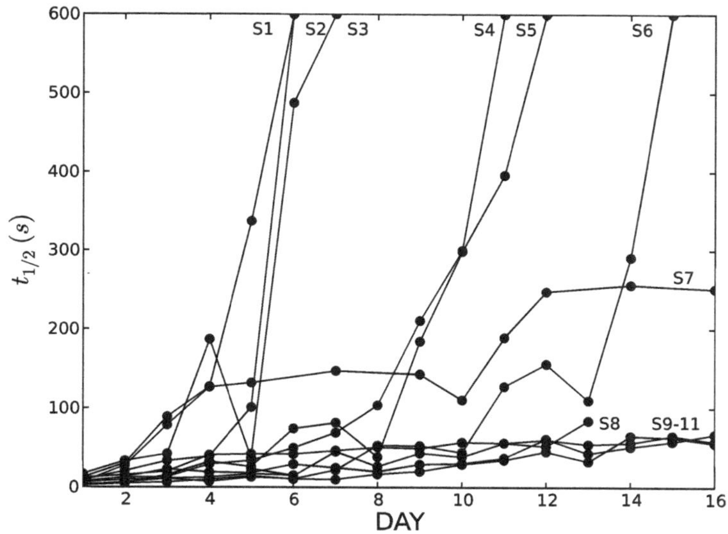

Fig. 9.4 Mean stick survival times, $t_{1/2}$, as a function of days of practice for 11 subjects

Skill Development

It has been shown that when time is rescaled by $t/t_{1/2}$, where $t_{1/2}$ is the mean stick survival time measured on a given day, the survival curves collapse onto a single survival curve (Cabrera and Milton 2012). This observation implies that $t_{1/2}$ is a relevant time scale for the development of stick balancing skill. Figure 9.4 plots $t_{1/2}$ as a function of days of practice. Each practice day consisted of the subject performing stick balancing as long as it takes to accumulate 10–15 min of total balance time (typically <1–2 h depending on the skill level of the subject). The time course for the development of skill, i.e., the increase in $t_{1/2}$, shows considerable variability between subjects. Although about 40–50 % of subjects became highly skilled ($t_{1/2} >$ 10 min), others achieved only moderate skill levels ($t_{1/2} <$ 2 min). Increases in $t_{1/2}$ with practice were not necessarily monotone: indeed those individuals who became most skilled often demonstrated a decrease in skill before the skill increased (e.g., S3 and S6). The observation that not all individuals become highly skilled despite practice is also seen in the acquisition of other complex voluntary motor tasks, such as ball throwing (Halverson et al. 1982).

It is important to note that in our laboratory the stick balancing task has purposely been made as difficult as possible to ensure that different skill levels can be readily identified. Thus, subjects are required to remain seated in a chair with their back against the back of the chair at all times. However, the fingertip, hand, and arm of the balancing limb are allowed to move freely in 3-D. An alternate strategy is to allow the subject to stand while balancing the stick at their fingertip (Cluff and Balasubramaniam 2009; Cluff et al. 2010, 2011). Subjects report that this version of stick balancing is easier and longer survival times are observed. However, key statistical properties, such as a Lévy distribution for the changes in speed made by the fingertip, are similar in both paradigms (compare Cabrera and Milton 2004a with Cluff and Balasubramaniam 2009).

Stabilization with Noise and Delay

Human stick balancing poses a similar challenge to neuroscientists as determining the mechanism for the flight of a bumblebee posed for aeronautical engineers (Wang 2000): it should be impossible, but nonetheless the organism performs very well! One possibility is that the nervous system employs some form of inverse model or predictive mechanism to anticipate the movements of the stick (see Discussion). However, a predictive mechanism capable of stabilizing in the presence of noise and delay has not yet been fully identified for stick balancing (Mehta and Schaal 2002). In this section and the next, we consider the possibility that the interplay between noise and delay makes the balancing task easier.

Postponement of Instability

The effects of the interplay between noise and delay on the value of a critical parameter were first studied in the context of the pupil light reflex modeled using the DDE (Longtin et al. 1990)

$$\dot{A}(t) + kA(t) = \frac{cK^n}{K^n + A^n(t-\tau)} + b \tag{9.10}$$

where b, c, K are constants, A is the pupil area, and reflex gain is proportional to n. The results are most easily understood by making reference to the bifurcation diagram for a supercritical Hopf bifurcation shown in Fig. 9.5a. The bifurcation occurs when $n = n_0$. The characteristic of this type of Hopf bifurcation is that there is an exchange in stability: the stable fixed-point becomes unstable; a stable limit cycle appears whose amplitude grows as $\sqrt{n - n_0}$.

Numerical simulations of Eq. 9.10 led to the prediction that noise shifted n_0 to a new position, $\hat{n}_0$, located to the right by an amount that was proportional to the variance, σ^2, of the noise and inversely proportional to its correlation time (Fig. 9.5b). This effect was seen whether the noise was multiplicative

$$c(t) = c_0 + \sigma^2 \xi(t)$$

or additive

$$b(t) = b_0 + \sigma^2 \xi(t)$$

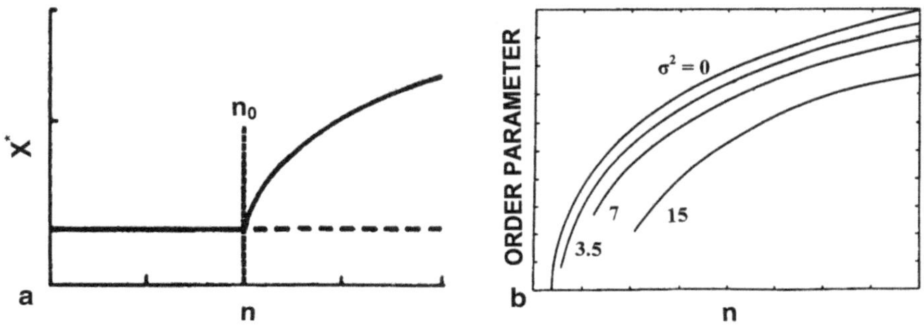

Fig. 9.5 **a** Schematic representation of a supercritical Hopf bifurcation. The *solid lines* represent stable solutions and the *dashed lines* the unstable solutions. In this diagram multiplicative noise produces horizontal transitions and additive noise produces transitions in the vertical direction. **b** Postponement of the bifurcation for Eq. 9.10 as a function of the intensity σ^2 of multiplicative noise. The order parameter is the distance between the two maxima in the probability density that occurs when the fixed-point become unstable. Similar results are obtained for additive noise. (Longtin et al. 1990)

where c_0, k_0 are positive constants and $\xi(t)$ describes exponentially correlated Gaussian distributed noise.

Although Eq. 9.10 successfully predicted the nature of the amplitude and period fluctuations at oscillation onset for the pupil light reflex, it was not possible to directly confirm the postponement of instability experimentally. In the presence of noise, the occurrence of a Hopf bifurcation is identified when the unimodal stationary probability density characteristic of a noisy fixed-point is replaced by a bimodal probability density whose two maxima separate as n increases. It was practically impossible to construct these densities given the short length of the available time series.

Recently it has been shown that a postponement of instability related to the effects of multiplicative noise occurs in a linearized model of an inverted pendulum with time-delayed feedback (Ushida 2011)

$$\ddot{\theta}(t) + \Gamma\dot{\theta}(t) + q\theta(t) = G(\xi(t))\theta(t-\tau) \tag{9.11}$$

where θ is the vertical displacement angle, $\Gamma \equiv 3k_d/m, q \equiv 3g/\ell, m$ is the stick mass and k_d is the damping coefficient. In this model, multiplicative noise enters through the gain where

$$G(\xi(t)) = G_0 + \sigma^2\xi(t)$$

and $\xi(t)$ is Gaussian-distributed white, i.e., delta-correlated, noise whose mean and variance are respectively, 0 and σ^2. It should be noted that in contrast to Eq. 9.10, a stable solution does not exist for Eq. 9.11 once the fixed point becomes unstable. It was shown that for appropriate choices of G_0 and q, the instability was postponed. In other words, for a given value of G_0, the value of q which destabilized the upright

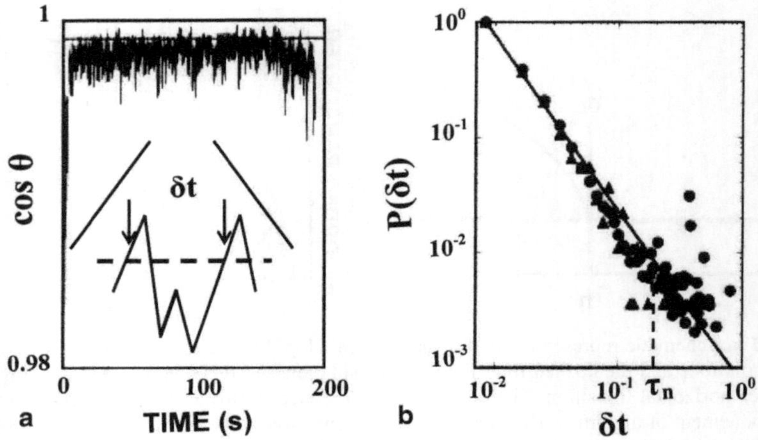

Fig. 9.6 **a** Temporal series of cosθ for a 0.62 m stick balanced at the fingertip. The inset defines how the laminar phases are estimated: δt refers to the time between successive crossings of a threshold (*dotted*) line in the upward (*corrective*) direction. **b** Normalized laminar phase probability distribution $P(\delta t)$. The *solid line* represents a power law with exponent $-3/2$ and the *vertical dashed line* indicates $\tau_n \sim 0.225$ ms

position was larger (the length of the stick was shorter). This prediction was confirmed experimentally using a seesaw-cart paradigm (Ushida 2011). Moreover, it was found that near the noisy stability boundary, $\theta(t)$ more rapidly converged to the upright position suggesting that maneuverability was increased

On–Off Intermittency

The first experimental evidence that multiplicative noise played an important role in human stick balancing was obtained by measuring the fluctuations in θ for 1-D (Foo et al. 2000; Jirsa et al. 2000) and for 3-D (Cabrera and Milton 2002). Figure 9.6a shows a plot of cos θ measured in 3-D using high speed motion capture cameras as a function of time. Since $\cos\theta = \Delta z/\ell$, $\cos\theta$ is a measure of the difference Δz in the vertical positions of the top and bottom of the stick. As can be seen, θ varies irregularly and, in particular, its average value is not precisely vertical (cos $\theta = 1$), but is displaced slightly from vertical. In physical sciences, a time series which undergoes rapid and irregular changes in amplitude raises the issue of intermittency. To test this possibility, it is necessary to measure the statistical properties of the laminar phases, namely, the length of the time interval, δt, between successive corrective movements. A corrective movement is defined as a movement which crosses a threshold in the upward (corrective) direction (see insert in Fig. 9.6a). Figure 9.6b shows that a log–log plot of the probability, $P(\delta t)$, of a laminar phase of length δt versus δt is linear with slope −3/2. Surprisingly, ≥98 % of the δt are shorter than estimates of τ_n for stick balancing (Fig. 9.2).

It was found that a −3/2 power law could be generated by slightly modifying Eq. 9.11 to

$$\ddot{\theta}(t) + \Gamma\dot{\theta}(t) + q \sin\theta(t) = G(\xi(t))\theta(t - \tau) \tag{9.12}$$

and adjusting the parameters to tune control sufficiently close to the edge of stability so that the feedback gain can be stochastically forced back and forth across the boundary (Cabrera and Milton 2002). This mechanism for generating a −3/2 power law is called *on–off intermittency*. It should be noted that for sufficiently small θ, we have $\sin\theta \approx \theta$. Thus the fact that Eq. 9.12 generates a power law, but Eq. 9.11 does not, emphasizes the importance of the larger θ for generating a power law.

The above observations suggest that in the presence of noise, the upright position of a stick can become statistically stabilized (Cabrera and Milton 2002). This is because the fluctuations in θ resemble a random walk whose mean value is approximately zero. Subsequently the importance of the initial function, Φ, for transiently stabilizing an unstable fixed-point was recognized through an analysis of the simpler DDE with additive noise (Milton et al. 2009c)

$$\dot{x}(t) - kx(t - \tau) = \sigma^2\xi(t) \tag{9.13}$$

where $k > 0$. When $\sigma^2 = 0$, the fixed-point is unstable. Suppose we chose $x(t_0) = 0$ and the initial function so that $\Phi(s) < 0$, for all $s \in [-\tau, 0)$. The effect of $\Phi(s)$ will be to transiently delay the escape of $x(t) \to +\infty$. This observation can be extended to include other choices of $\Phi(s)$. This phenomenon also occurs if Eq. 9.13 is represented as a delayed random walk (Milton et al. 2008). The point is that memory of the past can transiently act as a kind of negative feedback that pulls the system back towards the fixed-point.

An interesting historical note concerns the question as to whether the control system is tuned to the stable side of the stability boundary or to the unstable side. In the original numerical studies by Cabrera and Milton (2002), it was assumed that the control system was tuned slightly to the stable side of the stability boundary for the deterministic ("noise-free") system. This choice was made to simplify the numerical simulations. However, subsequent observations indicated that the control system was most likely tuned to the unstable side of the stability boundary (Cabrera et al. 2006). The observation that the stochastic stability boundary is located to the right of the deterministic stability boundary is consistent with the recently demonstrated noise-induced postponement of instability (Ushida 2011) discussed in the previous section.

Drift and Act Control

The existence of a power law for stick balancing does not readily identify the nature of the underlying neural control mechanisms. Indeed a −3/2 power law can be generated by a large number of different mechanisms (for a useful reviews see Sornette

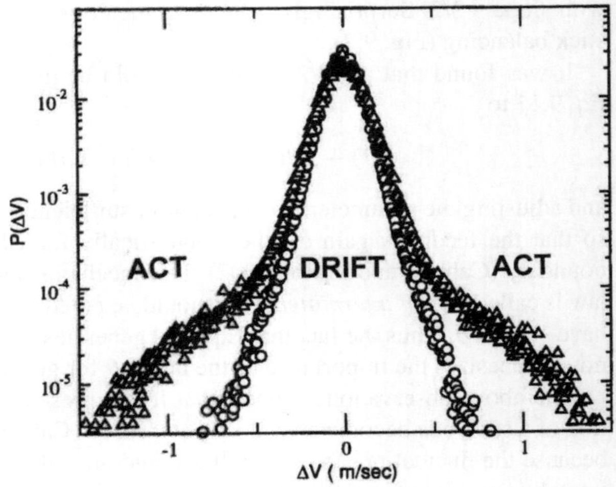

Fig. 9.7 Changes in the probability distribution, $P(\Delta V)$, for the changes in the speed of the fingertip, ΔV, as a function of expertise for stick balancing (Cabrera and Milton 2004). The stick length is 0.62 m. The distribution (o) was measured on the first day of practice and the distribution (Δ) was measured after 10 days of practice

2004, Newman 2005) and, in particular, are characteristic of adaptive control strategies (Eurich and Pawelzik 2005; Patzelt and Pawlezik 2011). Here we suggest that some form of a "drift and act" control strategy underlies these power law behaviors. The "drift and act" hypothesis posits that active corrections ("act") are made only when fluctuations exceed a threshold. For small fluctuations ("drift"), the control is provided through intrinsic mechanisms which include biomechanical properties and the interplay between noise and delay (Milton et al. 2008, 2009a, 2009b, 2009c).

A drift and act hypothesis is suggested by considerations of the probability distribution, $P(\Delta V)$, for the changes in speed, ΔV, made by the fingertip during stick balancing: the central portion of $P(\Delta V)$ for the more skilled stick balancer represents "drift" and the broad shoulders represent "act" (Fig. 9.7). As skill increases, $P(\Delta V)$ develops broad shoulders characteristic of a Lévy distribution (Cabrera and Milton 2004a; Cluff and Balasubramaniam 2009). Lévy-type distributions are characteristic of intermittent movement patterns ranging from the movements of microglial cells in brain slices (Grinberg et al. 2011) to *Drosophila* flight patterns (Reynolds and Frye 2007). This interpretation reflects the fact that intermittent dynamics can arise from the interaction between two opposing forces: a destabilizing and a delayed stabilizing force (Cabrera 2005).

Nested Control Loops

A possible realization of this mechanism would be a fixed-point whose basin of attraction is of the same order of size as the magnitude of the random perturbations that it receives (Milton et al. 2009a). In this setting, stability would be increased by constructing a "safety net" so that whenever trajectories escape the basin of attraction, they are redirected back inside (Cabrera and Milton 2012; Guckenheimer 1995).

Physiological control typically involves multiple feedback loops, each of which is associated with a different time delay and sensory threshold (Glass et al. 1988; Glass and Malta 1990). For example, the visual threshold for detecting a change in θ for stick balancing is likely to be $<1°$ (Orban et al. 1984) and $\tau \sim 225$ ms. On the other hand, the threshold for mechanoreceptors to detect a change in θ has been estimated to be 4–5° (Dodson et al. 1998), whereas the time delay is likely to be <100 ms (Mizobuchi et al. 2000).

These observations suggest the possibility of a nested control network topology in which the number of feedback loops that participate in control increases as the magnitude of the deviation from the set point to be controlled increases (Cabrera and Milton 2012). An example suggestive of a nested control network structure is the ankle-hip-step strategy used by humans to maintain balance in the face of increasingly larger perturbations (Shumway-Cook and Woollacott 2001). It has been hypothesized that chaotic dynamics can appear when there are three or more nested time-delayed feedback loops (Glass and Malta 1990).

An example of a simple nested control loop strategy for human balance control is

$$\dot{x}(t) = \begin{cases} kx(t) + \sigma^2\xi(t) + C & if \quad x(t-\tau) < -1 \\ kx(t) + \sigma^2\xi(t) & if \quad -1 \le x(t-\tau) \le 1 \\ kx(t) + \sigma^2\xi(t) - C & if \quad x(t-\tau) > 1 \end{cases} \tag{9.14}$$

where C is a positive constant and the feedback is represented by the switch-like discontinuous function shown in Fig. 9.8a (Eurich and Milton 1996). For small displacements, $|x| < 1$, the feedback is not active. Thus, the balance control system drifts under the influence of noise (open-loop control). This situation could reflect the presence of a sensory "dead zone", namely the displacements are too small to be detected by, for example, joint mechanoreceptors. When $|x| < 1$, corrective movements act to stabilize the system by reducing $|x|$ (closed-loop control).

In this simple model, the dynamics depend on only three parameters: the noise intensity (σ^2), the time delay (τ) and the restoring force (C). Not surprisingly, when $\sigma^2 = 0$, the behavior of Eq. 9.14 can be completely determined analytically (Eurich and Milton 1996). There is either a single limit cycle or the co-existence of two different limit cycles (multistability). Similar observations are obtained when Eq. 9.14 is modified to describe two nested feedback control loops (Milton et al. 2009a)

$$\dot{x}(t) = \begin{cases} kx(t-\tau) + \sigma^2\xi(t) + C & if \quad x(t-\tau') < -1 \\ kx(t-\tau) + \sigma^2\xi(t) & if \quad -1 \le x(t-\tau') \le 1 \\ kx(t-\tau) + \sigma^2\xi(t) - C & if \quad x(t-\tau') > 1 \end{cases} \tag{9.15}$$

where τ, τ' are two time delays that are not necessarily equal and $k > 0$. This formulation can potentially take advantage of the interplay between noise and delay for stabilizing the unstable fixed-point that is expected to occur for small k (Milton et al. 2008). Numerical simulations when $\tau = \tau'$ suggest that the behaviors generated by Eqs. 9.14 and 9.15 are very similar. Recently these concepts have been extended to incorporate a discontinuous PD-type controller, i.e., feedback that depends on both $x(t)$ and (t) (Kowalczyk et al. 2011). The fact that limit cycle oscillations and

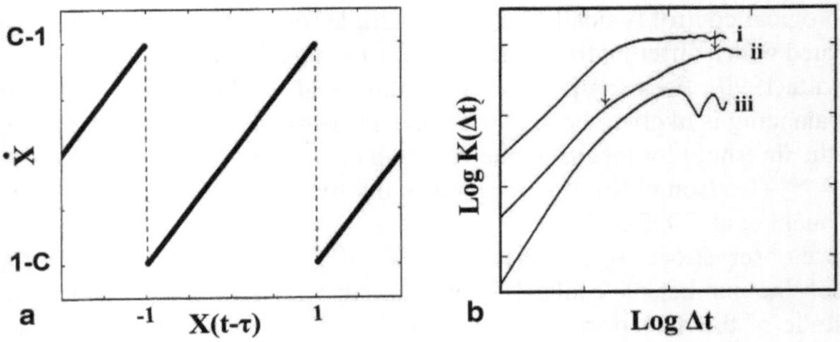

Fig. 9.8 **a** Schematic representation of the feedback described by Eq. 9.14. **b** Two-point correlation function for the fluctuations generated by Eq. 9.15. The noise intensity (σ^2) is highest for (i) and lowest for (iii). (The down arrows subdivide $K(\Delta t)$ into three regions, each of which can be approximated with a linear slope. For more details see Milton et al. 2009c.)

multistability also arise emphasizes the robustness of the occurrence of these dynamics in a nested control loop topology composed of only two time-delayed feedback controllers.

The basic conclusions of Eqs. 9.14 and 9.15 are supported by observations made on human postural sway. Postural sway refers to the fluctuations observed in the center of pressure while a subject stands quietly on a force platform with eyes closed. There is an underlying oscillatory trend to these fluctuations (Winter et al. 1998; Yamada 1995). An underlying oscillatory trend has also been emphasized for the movements of the fingertip in the horizontal plane for stick balancing (Milton et al. 2009b). Moreover, as illustrated in Fig. 9.8b, the two-point correlation function

$$K(\Delta t) = \frac{1}{N-m} \sum_{i=1}^{N-m} [(x(t_i) - x(t_i + \Delta t))^2 + (y(t_i) - y(t_i + \Delta t))^2]$$

where $\Delta t = t_1 - t_2$ exhibits a varying number of scaling regions (Collins and De Luca 1994; Milton et al. 2009b). It was observed that multiple scaling regions in $K(\Delta t)$ could be reproduced by tuning Eqs. 9.14 or 9.15 into the multistable regime and then using noise to cause switching between the two limit attractors. All of the patterns shown in Fig. 9.8b can be obtained simply by changing the noise intensity σ^2 (Milton et al. 2009c).

Switch-like Control of Saddle Points

Recently a novel suggestion concerning the role played by switch-like controllers in the stabilization of an inverted pendulum has been advanced (Bottaro et al. 2008; Asai et al. 2009). To illustrate this idea, consider the dynamics of an inverted planar pendulum linearized about the unstable upright fixed-point

$$\ddot{\theta}(t) - \frac{g}{\ell}\theta(t) = 0.$$

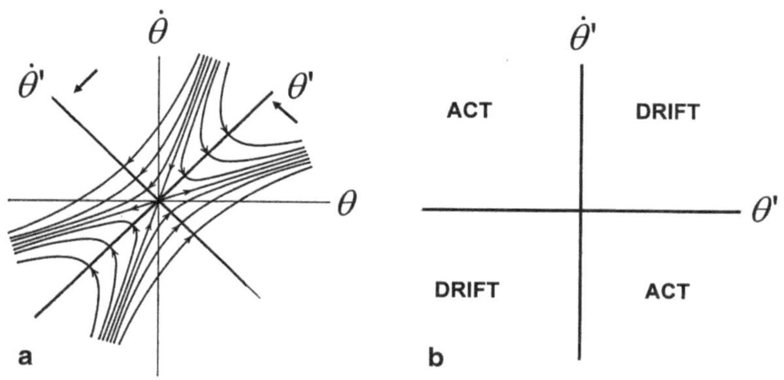

Fig. 9.9 **a** Phase plane representation of a saddle point. The axis are rotated to θ' vs. θ'. **b** The quadrants labeled "ACT" are when the control is switched on; otherwise the control is off ("DRIFT"; see text for discussion)

By making the Ansatz, $\theta(t) \approx e^{\lambda t}$, we see that $\lambda_{1,2} = \pm\sqrt{g/\ell}$ and thus the fixed-point is an unstable saddle point. Figure 9.9a shows a phase plane representation of a saddle point.

Each trajectory in the phase plane represents the solution for a different choice of the initial condition $(\theta(t_0), \dot{\theta}(t))$. If we look carefully at each trajectory we can see that even though the fixed-point is unstable there are intervals during which the trajectories approach the fixed-point and other intervals where they diverge from the fixed-point. There are only two exceptions to this rule: both correspond to a very special choice of the initial conditions and hence for all practical purposes can be ignored.

Figure 9.9b shows a switch-like control strategy that takes advantage of the phase plane properties of a saddle point. By rotating the axis of the phase plane from (Fig. 9.9a) we can see that the phase plane can be divided into regions where the trajectories approach the saddle point and other regions where the trajectories diverge away. Thus the investigators turned the control off ("drift") when the trajectories were in regions that naturally moved towards the fixed-point and activated control ("act") in the regions where the trajectories would naturally move away from the fixed-point (Fig. 9.9b). Since this switch-like controller depends on both $\theta(t)$ and $\dot{\theta}(t)$ it is a PD-type controller, albeit a more complicated one than the PD controllers typically used in engineering applications.

There were three surprising observations. First, it was possible to determine an act mechanism that resulted in robust control. Second, this strategy worked in the presence of a delay and hence for a saddle point that exists in an infinite dimensional space. Finally, in the presence of additive noise this control strategy results in the presence of power law scaling regions similar to those observed for human postural sway (Asai et al. 2009).

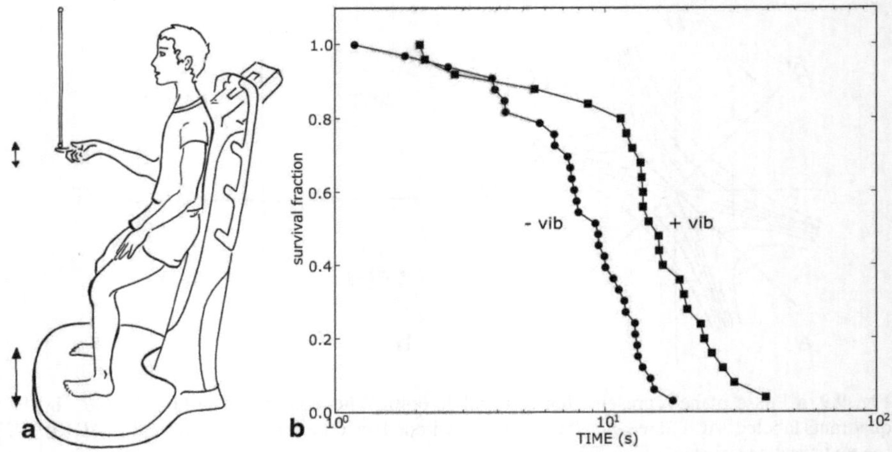

Fig. 9.10 **a** Schematic representation of stick balancing at the fingertip while standing on a vibrating platform. **b** Stick survival curves in the presence and absence of vibration for the same subject. (Figures reproduced from Milton et al. 2009a)

The Vibration Paradox

The introduction of vertical vibration at the fingertip using a whole-body vibrator benefits stick balancing (Milton et al. 2009a, Fig. 9.10a). In this experiment, a 3–5 mm vertical amplitude, 15–50 Hz vibration applied to the sole of the feet (Fig. 9.10a) produces a ~ 0.1 mm vertical amplitude vibration at the fingertip (the body behaves as a second-order low-pass filter with respect to vibration frequencies in this range). Figure 9.10b shows that the stick survival curve in the presence of vibration is shifted to the right, that is, $t_{1/2}$ is larger in the presence of vibration.

The result shown in Fig. 9.10b is unexpected. It is known that an inverted pendulum can be stabilized by moving the pivot point vertically either periodically (Acheson 1997) or noisily (Bogdanoff 1962; Bogdanoff and Citron 1964). These stabilizing effects are typically interpreted in the context of the Mathieu equation

$$\ddot{\theta}(t) + (k + b\cos 2\pi f_V t)\theta(t) = 0 \tag{9.16}$$

where b, k are positive constants. The condition for stability is that

$$f_V > \frac{\sqrt{2g\ell}}{2\pi h} \tag{9.17}$$

where h is the peak-to-peak amplitude. This condition ensures that the downward acceleration can exceed that of gravity (Pippard 1987). Thus the observations in Fig. 9.10 are surprising for two reasons. First, the downward acceleration cannot exceed g since the stick is not physically attached to the fingertip. Second, the vibration frequency that benefits stick balancing (50 Hz) is nearly 100 times smaller than that predicted by Eq. 9.17 (when $\ell = 0.55$ m and h $= 0.001$ m, then $f_V = 5525$ Hz). Subsequently, it was found that the application of low amplitude, low frequency vibration to the Achilles tendon had a stabilizing influence on human postural sway (Milton et al. 2010). How is this possible?

There have been two approaches to resolve this paradox. The first approach was to evaluate the stability of a Mathieu-type equation with time-delayed PD type feedback (Insperger 2011),

$$\ddot{\theta}(t) + (k + b\cos 2\pi f_V t)\theta(t) = -k_p\theta(t-\tau) - k_q\dot{\theta}(t-\tau) \tag{9.18}$$

where k_p, k_q are positive constants. Define the critical length of the stick as the stick length that can just be balanced against a given delay. It was found that Eq. 9.18 predicts that a shorter stick can be balanced than predicted by Eq. 9.16 and that this effect can occur even if the maximum acceleration of the stick's base does not exceed g. However, these effects require considerably larger vibration amplitude than used for human stick balancing.

The second approach has been to interpret the effects of vibration in the context of a "drift and act" hypothesis (Milton et al. 2009a). This mechanism proposes that the basin of attraction for the stabilized upright position is small enough so that escapes ("falls") are possible. Inside the basin of attraction, the trajectories "drift"; and corrective actions are taken only when trajectories leave the basin of attraction. Consequently any strategy that decreases the amplitude of the fluctuations in $\theta(t)$ will have a stabilizing effect since they decrease the probability that the trajectory escapes the basin of attraction. The main features of these observations can be qualitatively captured by a simple model which incorporates an unstable equilibrium point, a time-delayed switch-type controller and parametric periodic excitation

$$\dot{x}(t) = F(x(t-\tau))x(t) + kx(t)\sin 2\pi ft \tag{9.19}$$

where $F(x(t-\tau))$ is given in Fig. 9.11a. Equation 9.19 describes a "drift and act" controller: corrective actions ("act") are taken only when $Th_2 > x(t-\tau) > Th_2$. When $Th_2 >> Th_1$ there is a range of parameters for which a complex periodic attractor exists (Fig. 9.11b, 9.11c). The predicted decrease in the amplitude of the fluctuations and the shift in the position of the centroid have been observed in the fluctuations recorded for human postural sway when the Achilles' tendons are vibrated with low amplitude ($\sim$0.001 m) periodic vibration (Milton et al. 2010). If Th_2 is carefully adjusted so that escapes are possible in the presence of low intensity additive noise, then it is observed that periodic stimulation results in a prolongation of the survival time (Fig. 9.11d).

Concluding Remarks

The interplay between noise and delay can benefit the voluntary control of balancing tasks, such as stick balancing at the fingertip, in at least two ways. First, this interplay can postpone instability. This effect would be anticipated to facilitate control in the

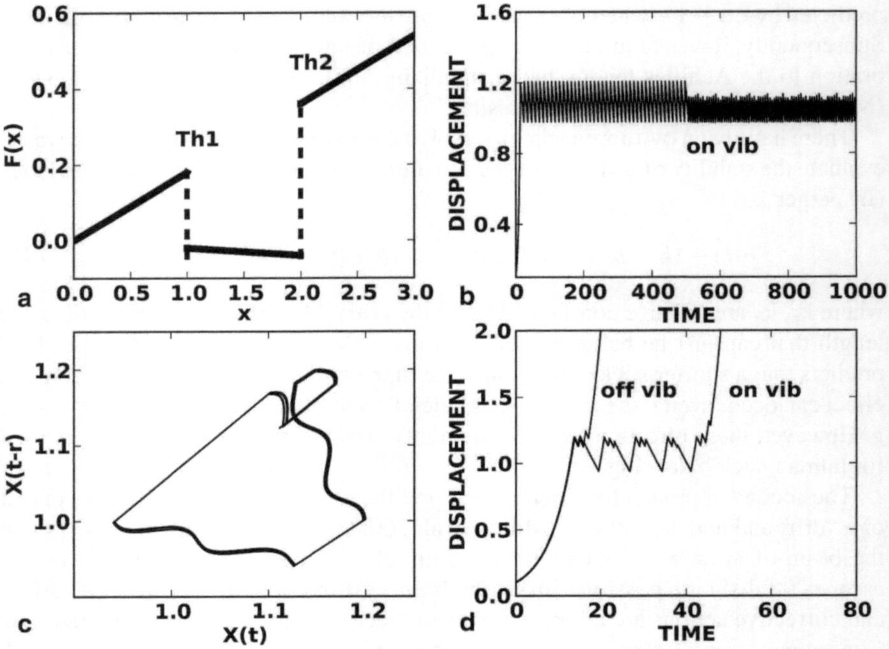

Fig. 9.11 **a** Graphical representation of $F(x(t-\tau))$ (for details see Milton et al. 2009c). **b** When $Th_2 >> Th_1$, periodic forcing decreases both the peak-to-peak and mean amplitude of the oscillation. **c** A complex attractor exists under these conditions. It is clearly more complex than a limit cycle; however, its exact nature has not yet been characterized. **d** When $Th_2 > Th_1$, the system can only be transiently confined and eventually escapes to infinity. Turning on the periodic forcing approximately doubles the survival time

sense that stable dynamics are easier to manage than unstable dynamics. Second, this interplay necessitates "drift and act" type control strategies. In addition, to stick balancing, intermittent corrective movements appear to be a component of the control of postural balance (Loram et al. 2005), saccadic eye movements, and the dynamics of cochlear hair cells (Moreau and Sontag 2003). Switch-like controllers are well known to engineers and have the property that they are optimal when the control is bounded (Flügge-Lotz 1968). Recently, it has been demonstrated that intermittent control of a joystick is better than continuous control in the performance of a virtual tracking task (Loram et al. 2011).

The above observations are to be contrasted with current hypotheses for voluntary motor control which portray time delays an obstacle for neural control. Indeed, it has been frequently argued that the nervous system must utilize mechanisms which anticipate the deleterious effects of delays (for a review see Nijhawan 2008). The proposed anticipatory mechanisms, in turn, emphasize the existence of internal models developed and used by the nervous system to base predictive actions (for a review see Shadmehr et al. 2010). However, neural populations that underlie these schemes are themselves time delayed. The observations presented here suggest that a more parsimonious view is to consider that noise and delay are part of the solution, not part of the problem.

It is perilous to develop models of neural control without considering the effects of noise and delay. In general it is not possible to reduce an infinite dimensional dynamical system to a finite dimensional system without losing information, except under certain very specific conditions (Insperger 2007). A fundamental problem is that the mathematical techniques most commonly employed in engineering control theory, namely the Fourier integral transform and Kalman filters, are not well suited for the study of the dynamics of transients and they are insensitive to the effects of the initial function, $\Phi(s)$. As approaches to movement science which combine mathematical modeling continue to develop, we can expect new theories for the control of voluntary movements by the nervous system to emerge (Milton 2009).

An unresolved question is whether or not models of balance control can fully capture the dynamics of stick balancing. This is because there are two distinct mechanisms that can cause the stick to fall: (1) θ becomes too large, and (2) the required movement of the fingertip to keep the stick balanced lies outside the reach of the arm (in our laboratory the subject's back must be kept against the back of a chair at all times). Thus not all movements of the fingertip are directed towards minimizing θ: some movements must be made to maintain the balancing task within arm's reach. An alternate view is that stick balancing has characteristics of a delayed pursuit-escape task in which the fingertip pursues the tip of the stick (Milton et al. 2011; Ohira et al. 2011).

A characteristic of living organisms is the intermittency in their movement activities. Well studied examples include human movements (Gross et al. 2002), swimming patterns of fish (Viswanathan 2010), the flights of fruit flies (Reynolds and Frye 2007) and birds (Rayner et al. 2001), and foraging patterns (for a review see Viswanathan 1999). Intermittent fluctuations can also be observed in the collective activity of populations (Cole 1991) and even in traffic flow along axons (Roy et al. 2000) and highways (Sugihara et al. 2008). This intermittency may also be interpreted as segments in which the movements drift in response to the environment with little active control interspersed with relatively shorter segments in which the organism acts by making directed movement corrections. It is interesting to speculate that some of these phenomena might also reflect a drift and act control strategy.

Acknowledgments The author acknowledges useful discussions with J. L. Cabrera, S. A. Campbell, T. Insperger, A. Longtin, M. C. Mackey, T. Ohira, R. Peterka, G. Stépán, and A. Ruina. The author was supported by the William R Kenan, Jr., Foundation and the National Science Foundation (NSF-1028970).

References

Acheson D (1997) From calculus to chaos: an introduction to dynamics. Oxford University Press, New York

Areili A, Sterkin A, Grinald A, Aertson A (1996) Dynamics of ongoing activity: explanation for the large variability in evoked potential responses. Science 273:1868–1871

Asai Y, Tasaka Y, Nomura K, Nomura T, Casidio M, Morasso P (2009) A model of postural control in quiet standing: robust compensation of delay-induced instability using intermittent activation of feedback control. PLoS ONE 4:e6169

Beilock S (2011) Choke: what the secrets of the brain reveal about getting it right when you have to. Simon & Schuster: Free Press, New York

Berger RD, Saul JP, Cohen RJ (1989) Transfer function analysis of autonomic regulation. I. Canine atrial rate function. Am J Physiol 256:H142–H152

Bogdanoff JL (1962) Influence on the behavior of a linear dynamical system of some imposed rapid motions of small amplitude. J Acoust Soc Amer 34:1055–1062

Bogdanoff JL, Citron SJ (1964) Experiments with an inverted pendulum subject to rapid parametric excitation. J Acoust Soc Amer 38:447–452

Bormann R, Cabrera JL, Milton JG, Eurich CW (2004) Visuomotor tracking on a computer screen—an experimental paradigm to study the dynamics of motor control. Neurocomp 58–60:517–523

Bottaro A, Yasutake Y, Nomura T, Casidio M, Morasso P (2008) Bounded stability of the quite standing posture: An intermittent control model. Hum Mov Sci 27:473–495

Bressloff PC, Wood CV (1997) Spontaneous oscillation in a nonlinear delay-feedback shunting model of the pupil light reflex. Phys Rev E 58:3597–3606

Bressloff PC, Wood CV, Howarth PA (1996) Nonlinear shunting model of the pupil light reflex. Proc Roy Soc B 263:953–960

Cabrera JL (2005) Controlling instability with delayed antagonist stochastic dynamics. Physica A 356:25–30

Cabrera JL, Milton JG (2002) On-off intermittency in a human balancing task. Phys Rev Lett 89:158702

Cabrera JL, Milton JG (2004a) Human stick balancing: tuning Lévy flights to improve balance control. Chaos 14:691–698

Cabrera JL, Milton JG (2004b) On-off intermittency and survival times. Nonlinear Studies 11:305–317

Cabrera JL, Milton JG (2012) Stick balancing, falls and dragon kings. Eur J Phys Special Topics 205:231–241

Cabrera JL, Bormann R, Eurich CW, Ohira T, Milton J (2004) State-dependent noise and human balance control. Fluct Noise Lett 4:L107–L117

Cabrera JL, Luciano C, Milton J (2006) Neural control on multiple times scales: insights from human stick balancing. Condens Matter Phys 9:373–383

Chiel HJ, Beer RD (1997) The brain has a body: adaptive behavior emerges from interactions of nervous system, body and environment. TINS 20:553–557

Cluff T, Balasubramaniam R (2009) Motor learning characterized by changing Lévy distributions. PLoS One 4:e5998

Cluff T, Gharib T, Balasubramaniam R (2010) Attentional influences on the performance of secondary physical tasks during balance control. Exp Brain Res 203:647–658

Cluff T, Boulet J, Balasubramaniam R (2011) Learning a stick balancing task involves task-specific coupling between posture and hand displacements. Exp Brain Res 213:15–25

Cole BJ (1991) Short-term activity cycles in ants: generation of periodicity by worker interactions. Amer Natur 137:244–259

Collins JJ, De Luca CJ (1994) Random walking during quiet standing. Phys Rev Lett 73:764–767

Crutchfield JP, Farmer JD, Huberman BA (1982) Fluctuations and chaotic dynamics. Phys Rep 92:45–82

Dodson MJ, Goodwin AW, Browning AS, Gehring HM (1998) Peripheral neural mechanisms determining the orientation of cylinders grasped by the digits. J Neurosci 18:521–530

Eurich CW, Milton JG (1996) Noise-induced transitions in human postural sway. Phys Rev E 54:6681–6684

Eurich CW, Pawelzik K (2005) Optimal control yields power law behavior. In: Duch W, Kacprzyk J, Oja E, Zadronzny S Lecture notes in computer science, No. 3697, Springer-Verlag, Berlin, pp 365–370

Fatt P, Katz B (1950) Some observations on biological noise. Nature 166:597–598

Flügge-Lotz I (1968) Discontinuous and optimal control. McGraw-Hill, New York

Foo P, Kelso JAS, de Guzman GC (2000) Functional stabilization of fixed points: human pole balancing using time to balance information. J Exp Psychol: Hum Percept Perform 26:1281–1297

Freeman WJ, Holmes MD, West GA, Vanhatalo S (2006) Dynamics of human neocortex that optimizes its stability and flexibility. Int J Intell Sys 21:881–901

Gabbard CP (2004) Lifelong motor development, 4th edn. Benjamin Cummings, San Francisco

Gawthrop P, Loram I, Lakie M, Gollee GH (2011) Intermittent control: a computational theory of human control. Biol Cybern 104:31–51

Glass L, Malta CP (1990) Chaos in multilooped negative feedback loops. J Theoret Biol 145:217–223

Glass L, Beuter A, Larocque D (1988) Time delays, oscillations and chaos in physiological control systems. Math Biosci 90:111–125

Grinberg YY, Milton JG, Kraig RP (2011) Spreading depression sends microglia on Lévy flights. PLoS ONE 6:e19294

Gross J, Timmerman L, Kujala J, Dirks M, Schmitz F, Salmelin R, Schnitzler A (2002) The neural basis of intermittent motor control in humans. Proc Natl Acad Sci U S A 19:2299–2302

Guckenheimer J (1995) A robust hybrid stabilization strategy for equilibria. IEEE Trans. Automatic Control 40:321–326

Guckenheimer J, Holmes P (1983) Nonlinear oscillations, dynamical systems, and bifurcations of vector fields. Springer-Verlag, New York

Halverson LE, Robertson MA, Langendorfer S (1982) Development of the overarm throw: movement and ball velocity changes by seventh grade. Res Quart Exerc Sport 53:198–205

Harris CM, Wolpert DM (1998) Signal-dependent noise determines motor planning. Nature (London) 394:780–784

Hick WE (1952) On the rate of gain of information. Quart J Exp Psychol 4:11–26

Insperger T (2007) Act-and-wait concept for continuous-time control systems with feedback delay. IEEE Trans Control Sys Technol 14:974–977

Insperger T (2011) Stick balancing with reflex delay in case of parametric forcing. Commun Nonlinear Sci Simulat 16:2160–2168

Jirsa VK, Fink P, Foo P, Kelso JAS (2000) Parametric stabilization of biological coordination: a theoretical model. J Biol Phys 26:85–112

Kowalczyk P, Glenndinning P, Brown M, Medrano-Cerda G, Dallali H, Shapiro J (2011) Modelling human balance using switched systems with linear feedback control. J R Soc Interface 9:234–245

Kuiken TA, Miller LA, Lispchutz RD, Lock BA, Stubblefield K, Marasso PD, Zhou P, Dumanian G (2007) Targeted reinnervation for enhanced prosthetic arm function in a woman with a proximal amputation: a case study. Lancet 369:371–380

Linz SJ, Lücke M (1986) Effect of additive and multiplicative noise on the first bifurcation of the logistic model. Phys Rev A 33:2694–2703

Lockhart DB, Ting LH (2007) Optimal sensorimotor transformations for balance. Nature Neurosci 10:1329–1336

Longtin A, Milton JG (1989) Modeling autonomous oscillation in the human pupil light reflex using nonlinear delay-differential equations. Bull Math Biol 51:605–624

Longtin A, Milton JG, Bos JE, Mackey MC (1990) Noise and critical behavior of the pupil light reflex at oscillation onset. Phys Rev A 41:6992–7005

Loram ID, Maganaris CN, Lakie M (2005) Human postural sway results from frequent, ballistic bias impulses by soleus and gastrocnemius. J Physiol 564:295–311

Loram ID, Gollee H, Lakie M, Gawthrop PJ (2011) Human control of an inverted pendulum: is continuous control necessary? Is intermittent control effective? Is intermittent control physiological? J Physiol 589(2):307–324
MacDonald N (1989) Biological delay systems: linear stability theory. Cambridge University Press, New York
Mackey MC (1979) Periodic auto-immune hemolytic anemia: an induced dynamical disease. Bull Math Biol 41:829–834
Mackey MC, Glass L (1977) Oscillations and chaos in physiological control systems. Science 197:287–289
Mackey MC, an der Heiden U (1984) The dynamics of recurrent inhibition. J Math Biol 19:211–225
Matthews PBC (1996) Relationship of firing intervals of human motor units to the trajectory of post-spike after-hyperpolarization and synaptic noise. J Physiol 493(3):597–628
Mehta B, Schaal S (2002) Forward models in visuomotor control. J Neurophysiol 88:942–953
Milton JG (2009) Introduction to focus issue: bipedal locomotion—from robots to humans. Chaos 19:026101
Milton JG (2011) The delayed and noisy nervous system: implications for neural control. J Neural Engg 8:065005
Milton J, Small S, Solodkin A (2004) On the road to automatic: dynamic aspects of skill acquisition. J Clin Neurophysiol 21:134–143
Milton J, Cabrera JL, Ohira T (2008) Unstable dynamical systems: delays, noise and control. EPL 83:48001
Milton JG, Ohira T, Fraiser RM, Gyorffy JB, Ruiz FK, Strauss MA, Balch EC, Marin PJ, Alexander JL (2009a) Balancing with vibration: a prelude for "drift and act" balance control. PLoS ONE 4:e7427
Milton J, Cabrera JL, Ohira T, Tajima S, Tonosaki Y, Eurich CW, Campbell SA (2009b) The time-delayed pendulum: Implications for human balance control. Chaos 19:026110
Milton J, Townsend JL, King MA, Ohira T (2009c) Balancing with positive feedback: the case for discontinuous control. Phil Trans Roy Soc A 367:1181–1193
Milton J, Gyorffy J, Cabrera JL, Ohira T (2010) Amplitude control of human postural sway using Achilles tendon vibration. 16th US National Congress of Theoretical and Applied Mechanics, State College, PA, USNCTAM2010-791
Milton J, Lippai J, Bellows R, Blomberg A, Kamimura A, Ohira T (2011) Visuomotor tracking tasks with delayed pursuit and escape. ASME 2011 International Design Engineering Technical Conferences & Computers and Information in Engineering Conference, Washington, DC, DETC2011-47312
Mizobuchi K, Kuwabara S, Toma S, Nakajima Y, Ogawara K, Hattori T (2000) Single unit responses of human cutaneous mechanoreceptors to air-puff stimulation. Clin Neurophysiol 111:1577–1581
Moreau L, Sontag E (2003) Balancing at the border of stability. Phys Rev E 68:020901
Newman MEJ (2005) Power laws, Pareto distributions and Zipf's law. Contemp Phys 46:323–351
Nijhawan R (2008) Visual prediction: psychophysics and neurophysiology of compensation for time delays. Behav Brain Sci 31:179–239
Ohira T, Kamimura A, Milton JG (2011) Pursuit-escape with distance-dependent delay. 7th European Nonlinear Dynamics Conference (ENOC2011), Rome, Italy, MS-11
Orban GA, Vandenbussche E, Vogels R (1984) Human orientation discrimination tested with long stimuli. Vis Res 24:121–128
Patzelt F, Pawlezik K (2011) Criticality of adaptive control dynamics. Phys Rev Lett 107:238103
Pippard AB (1987) The inverted pendulum. Eur J Phys 8:203–206
Priplata A, Niemi J, Salen M, Harry J, Lipsitz LA, Collins JJ (2002) Noise-enhanced human balance control. Phys Rev Lett 89:238101
Rayner JMV, Viscardi PW, Ward S, Speakman JR (2001) Aerodynamics and energetics of intermittent flight in birds. Amer Zool 41:188–204

Reynolds AM, Frye MA (2007) Free-flight odor tracking in *Drosophila* is consistent with an optimal intermittent scale-free search. PLoS ONE 4:e354
Riani M, Simonotto E (1994) Stochastic resonance in the perceptual interpretation of ambiguous figures: a neural network model. Phys Rev Lett 72:3120–3123
Roy S, Coffee P, Smith G, Liem RKH, Brady ST, Black MM (2000) Neurofilaments are transported rapidly but intermittently in axons: implications for slow axonal transport. J Neurosci 20:6849–6861
Shadmehr R, Smith MA, Krakauer JW (2010) Error correction, sensory prediction and adaptation in motor control. Ann Rev Neurosci 33:89–108
Sieber J, Krauskopf B (2004a) Complex balancing motions of an inverted pendulum subject to delayed feedback control. Phys D 197:332–345
Sieber J, Krauskopf B (2004b) Bifurcation analysis of an inverted pendulum with delayed feedback control near a triple-zero eigenvalue singularity. Nonlinearity 17:85–103
Shumway-Cook A, Woollacott MH (2001) Motor control: theory and practical applications. Williams & Wilkins, New York
Sornette D (2004) Critical phenomena in natural sciences: chaos, fractals, self-organization and disorder: concepts and tools. Springer, New York
Stark L, Cornsweet TN (1958) Testing a servoanalytic study of consensual pupil oscillations. Science 127:588
Stepan G (2009) Delay effects in the human sensory system during balancing. Phil Trans Roy Soc A 367:1195–1212
Stépán G, Insperger T (2006) Stability of time-periodic and delayed systems—a route to act-and-wait control. Ann Rev Control 30:159–168
Stepp N (2009) Anticipation in feedback-delayed manual tracking of a chaotic oscillator. Exp Brain Res 198:521–525
Sugihara Y, Fukui M, Kikuchi M, Hasebe K, Nakayama A, Nishinari K, Tadaki S-I, Yukawa S (2008) Traffic jams without bottlenecks—experimental evidence for the physical mechanism of the formation of a jam. New J Phys 10:033001
Suminski AJ, Tkach DC, Fagg AH, Hatsopoulos NG (2010) Incorporating feedback from multiple sensory modalities enhances brain-machine interface control. J Neurosci 30:16777–16787
Todorov E (2004) Optimality principles in sensorimotor control. Nature Neurosci 7:907–915
Todorov E, Jordan MI (2002) Optimal feedback control as theory of motor coordination. Nature Neurosci 5:1226–1235
Ushida S (2011) Control performance improvements due to fluctuations in dynamics of stochastic control systems. 50th IEEE Conference on Decision and Control, Orlando, Florida, Dec 2011
Van Der Kooij H, van Asseldonk E, Van Der Helm FCT (2005) Comparison of different methods to identify and quantify balance control. J Neurosci Meth 145:175–203
Venkadesan M, Guckenheimer J, Valero-Cuevas F (2007) Manipulating the edge of stability. J Biomech 40:1653–1661
Viswanathan GM (1999) Optimizing the success of random searches. Nature (London) 401:911–914
Viswanathan GM (2010) Fish in Lévy-flight foraging. Nature (London) 465:1018–1019
Voss HU (2000) Anticipating chaotic synchronization. Phys Rev E 61:5115–5119
Wang ZJ (2000) Two dimensional mechanisms for insect hovering. Phys Rev Lett 85:2216–2219
Winter DA, Patla AE, Prince E, Ishac M, Gielo-Perczak K (1998) Stiffness control of balance during quiet standing. J Neurophysiol 80:1211–1221
Yamada N (1995) Chaotic swaying of the upright posture. Hum Mov Sci 14:711–726
Yildirim N, Santillán M, Horike D, Mackey MC (2004) Dynamics and bistability in a reduced model of the *lac* operon. Chaos 14:279–292

Index

M. J. Richardson et al. (eds.), *Progress in Motor Control*,
Advances in Experimental Medicine and Biology 782,
DOI 10.1007/978-1-4614-5465-6,

MIX
Papier aus verantwortungsvollen Quellen
Paper from responsible sources
FSC® C105338

If you have any concerns about our products,
you can contact us on
ProductSafety@springernature.com

In case Publisher is established outside the EU,
the EU authorized representative is:
Springer Nature Customer Service Center GmbH
Europaplatz 3, 69115 Heidelberg, Germany

Printed by Libri Plureos GmbH
in Hamburg, Germany